SpringerBriefs in Biochemistry and Molecular Biology

For further volumes:
http://www.springer.com/series/10196

David Escors · Karine Breckpot
Frederick Arce · Grazyna Kochan
Holly Stephenson

Lentiviral Vectors and Gene Therapy

Springer

David Escors
Division of Infection and Immunity
Rayne Institute
University College
University Street 5
London WC1E 6JF
UK

Karine Breckpot
Department of Physiology–Immunology
Medical School
Free University of Brussels
Laarbeeklaan 103
1090 Jette
Belgium

Frederick Arce
University College London
University Street 5
London WC1E 6JF
UK

Grazyna Kochan
Structural Genomics Consortium (SGC) Oxford
University of Oxford
Windmill Road
Oxford OX3 7HE
UK

Holly Stephenson
University College London
University Street 5
London WC1E 6JF
UK

ISSN 2211-9353 e-ISSN 2211-9361
ISBN 978-3-0348-0401-1 e-ISBN 978-3-0348-0402-8
DOI 10.1007/978-3-0348-0402-8
Springer Basel Heidelberg New York Dordrecht London

Library of Congress Control Number: 2012934664

Printed on acid-free paper

Springer is part of Springer Science+Business Media (www.springer.com)

Preface

Writing a book about lentivectors and gene therapy was certainly a challenge. During the past decades, retroviral and lentiviral vectors have moved from just academic and basic research into human clinical trials. Thus, the enormous amount of information makes it hard to select and distill those publications that made a real difference.

In addition to that, all this information has to be written to reach a broad range of readers, whether they are just beginners or fully accomplished scientists with a deep insight into the subject.

Here we have tried to accomplish just that. We asked ourselves the question of "what type of manual would I like my graduate/PhD students to read before starting to work in my lab?" Well, this work is the result of trying to answer that question. We think we are touching a wide variety of subjects, but with a focus on the final application of lentivectors into human gene therapy.

The Chap. 1 gives a general overview on the history of gene therapy, not necessarily involving lentivectors. Chapter 2 describes the development of retrovirus and lentivirus vectors, while in the Chaps. 3 and 4 their application mainly in experimental models. Chapters 4 and 5 are devoted to their application in human gene therapy.

We sincerely hope that you will enjoy reading this book, and that you may use it either as an "advanced" introduction into lentivectors and gene therapy, or as a useful and handy "manual" for you graduate students.

David Escors, Karine Breckpot, Frederick Arce,
Grazyna Kochan, and Holly Stephenson

Contents

Chapter 1
Introduction to Gene Therapy

David Escors and Karine Breckpot

Abstract Gene therapy as we understand it nowadays was conceived during the early and mid part of the twentieth century. At first, it was considered a revolutionary biomedical procedure, which could potentially cure any disease for which the molecular bases were understood. Since then, gene therapy has gone through many stages and has evolved from a nearly unrealistic perspective to a real life application. After several decades of research, a wide range of gene delivery vectors have been engineered and successfully tested in many animal models of human disease. However, clinical efficacy in humans could not be shown until the beginning of this century after its successful application in small-scale clinical trials to cure severe immunodeficiency in children. In these particular clinical trials, a retrovirus vector based on mouse leukemia virus was used, and their successes were overshadowed some time later by the occurrence of vector-related leukemia in a number of treated children. These fatal secondary effects clearly showed that the safe application of gene therapy critically depends on our understanding of vector engineering. In this context, lentiviral vectors have appeared, with improved efficiency and, apparently, increased biosafety. Very recently, the first clinical trials with lentivectors have been carried out with some success. In this chapter, we briefly define gene therapy, and describe the main scientific steps, which culminated in the engineering viral vectors in gene therapy, and place them in the context of current human therapy.

D. Escors (✉)
University College London, Rayne Building, 5 University Street, London WC1E 6JF, UK
e-mail: d.escors@ucl.ac.uk

K. Breckpot
Vrije Universiteit Brussels, Brussels, Belgium

D. Escors et al., *Lentiviral Vectors and Gene Therapy*,
SpringerBriefs in Biochemistry and Molecular Biology,
DOI: 10.1007/978-3-0348-0402-8_1, © The Author(s) 2012

1.1 The Concept of Gene Therapy

If we could come up with an appropriate definition of gene therapy, that would be the treatment of medical disorders by the delivery of therapeutic genes into the appropriate cellular targets. These therapeutic genes should correct deleterious effects from specific gene mutations in the appropriate cell types or tissues. Additionally, therapeutic genes can modify cell activities to overcome or prevent specific diseases, not necessarily with a genetic etiology. For successful gene therapy, the exogenous therapeutic gene has to be specifically, efficiently, and stably incorporated into the target cell. However, what do we consider a therapeutic gene? The definition of gene has been changing along with the increasing knowledge in molecular and cellular biology. Initially, it was a "bit of DNA" encoding a transmissible trait. Then, it was narrowed down to the DNA section that encodes an enzyme. After that, the definition was changed to include structural and enzymatic proteins. However, genes also encode several types of RNAs, namely ribosomal, transfer, and small nuclear RNAs. Even more, a new family of regulatory RNAs have been discovered, microRNAs, which adds a new level of "gene regulation". Therefore, the specific concept of gene therapy still remains somewhat elusive. Taking into account all these considerations, our original definition could be changed to the treatment of medical disorders by the delivery of therapeutic genetic information into the appropriate cellular targets.

1.2 Origins of Gene Therapy

It is very likely that the first thought coming to our readers' minds is that gene therapy is a cutting-edge, fairly recent biomedical technique, which can virtually cure any type of disease with a genetic cause. While part of this assumption might be true, gene therapy is by no means a novel biomedical concept. Quite on the contrary, the concept of gene therapy is not a recent one at all, and the idea of gene manipulation has always been controversial. At the end of the nineteenth and beginning of the twentieth century, the realization that traits had a genetic transmissible nature was taken out of context, leading to serious proposals that "bad traits" should be eliminated from the "human gene pool" (eugenesis) in a very misguided interpretation of Charles Darwin's natural selection theory. Since then, this and other misunderstandings of the concepts of genetics have encouraged controversy within the scientific and nonscientific community. Thus, to properly understand gene therapy and its derivation to the current therapeutic applications, it is necessary to put it in place from a historical point of view.

Gene therapy came about as the direct result of the extraordinarily rapid development of several scientific disciplines, such as medicine, molecular and cellular biology, and virology. In fact, the birth of gene therapy is to be found within a question that many have asked at one point or another in their lives.

Why do offspring look like their parents? Why are siblings so similar? From a qualitative, nonscientific point of view, it was fairly clear that there was some kind of transmission of particular traits, such as eye, hair color, particular facial features, or even diseases that run within specific families. However, the nature of these "transmissible traits" was very elusive, at least until the pioneering work of Gregor Mendel between 1850s and 1860s. What Mendel did extraordinarily right was to study simple qualitative traits, which could be easily tracked throughout generations. In addition, he chose as an experimental model mainly the garden pea plant, easy to grow, maintain, and study. He studied the frequency of qualitative traits such as flower and seed color or shape throughout many generations of crossings between different varieties. These experiments were groundbreaking, as Mendel demonstrated that transmission of qualitative traits followed specific mathematical rules. These transmission rules could only be explained if these hereditary characteristics existed as physically discreet entities, which were transmitted from parents to offspring. Ronal Fisher during the early twentieth century extended the findings by Mendel by applying mathematical models to quantitative traits [1]. Their combined work confirmed that living beings transmitted encoded information to offspring. This encoded information could be studied and its transmission could be accurately predicted following mathematical rules. The empirical biological material encoding genetic information was termed "gene" (from the greek generation) by the Danish botanist Wilhelm Johannsen in 1909 [2].

It has always been intuitively known that some human diseases such as hemophilia, β-thalassemia, particular types of cancer (familial retinoblastoma and colon cancer) and even diabetes run in particular families, pointing to genetic factors contributing to disease onset. These diseases were transmitted from parents to offspring more frequently in some families than they did in others without a medical history of the disease. In any case, early in the twentieth century, the correction of these diseases using gene therapy approaches was considered nearly an unsurpassable medical and scientific challenge. Genes could not be directly manipulated because their nature was unknown [3].

Between the 1940s and 1980s there was an unprecedented scientific explosion of breakthroughs in genetics, biochemistry, and molecular biology. Suddenly, gene therapy dramatically changed from an unrealistic possibility to a nearly certainty [4]. Firstly, DNA and not protein was identified as the biological molecule encoding genetic information [5]. DNA could be isolated and studied. A short time after that, its structure was solved [6, 7]. The DNA organization as a double antiparallel helix maintained by binding of specific pairs of deoxinucleotides provided a mechanistical explanation for the physical transmission of genetic information. More importantly, the isolation of the enzyme polynucleotide phosphorylase allowed the in vitro synthesis of RNA molecules of known nucleotide composition [8]. These synthetic RNA molecules permitted the deciphering of the genetic code, that is, the equivalence between nucleotide triads (codons) and specific aminoacids in a polypeptide chain [9]. Finally, gene cloning and gene delivery into mammalian cells was becoming routine in the 1970s and 1980s [10–15]. Thus, gene isolation and manipulation became a certainty.

1.3 Gene Therapy in the 1970s

It was by the late 1970s when gene therapy was seriously considered as a close realistic biomedical alternative. The molecular causes of several genetic diseases were well understood from previous biochemical research. Theoretically, gene therapy was an ideal and clean solution for the correction of at least some type of diseases such as hemophilia and β-thalassemia [16]. The concept was surprisingly simple, restore the gene, and thus cure the disease. Gene therapy was considered a solution for many human genetic conditions, and not an unrealistic alternative [16, 17].

In fact, it has been nearly forgotten that the first human gene therapy clinical trial was performed in the 1970s, instead of the 1980–1990s as it is common belief even within the scientific community. In this trial, three human patients suffering from hyperargininemia were intravenously injected with a semi-purified preparation of Shope papillomavirus [17, 18]. This rabbit papillomavirus was known to encode a viral arginase enzyme that corrected the disease in rabbit models of the disease. In fact, its intravenous administration in rabbits was asymptomatic, and no potential threats for human patients were contemplated. Surprisingly, at least at that time, no arginine reduction was detected in blood of treated patients. The authors explained the failure to the virus instability, and this particular therapeutic approach was not pursued further [17, 18]. However, this failed clinical trial tried to cover the gap between in vitro work and human application. As a consequence of the rapid scientific development, gene therapy became a real possibility, and invariably, it gave rise to moral and practical concerns. Curiously, even after 40 some years many of these concerns are very much the same, such as human cloning and genetic manipulation [19], especially after the advances in somatic cloning and regenerative medicine using embryonic and somatic stem cells.

1.4 Gene Therapy in the 1980s

During the 1980s, gene transfer methods to mammalian cells were developed and improved. Gene cloning and gene transfer was routinely performed. Although there were several viral and nonviral methods for gene transfer, the ones based on retrovirus vectors were clearly advantageous for gene therapy. Gene transfer vectors based on retroviruses such as mouse leukemia virus (MLV) were developed. Retroviruses are RNA viruses, which stably integrate their genome in the host cell chromosomes, so they were the ideal tool to introduce therapeutic genes [20]. Thanks to the "brand-new" molecular cloning techniques, the retrovirus genome was separated into at least a transfer vector plasmid and packaging plasmids encoding the structural/RT proteins [21]. Only the transfer vector contained the specific packaging signal that allowed its incorporation *in trans* by the structural proteins within virus-like particles. These particles were noninfectious and resulted in very efficient gene carriers. However, their therapeutic use for gene therapy was still rather controversial. There were concerns about the consequences of introducing exogenous DNA in cells, or even the lack of regulation of gene expression in the target cells, amongst other ethical issues [10, 22]. Some of these concerns were brought alight when an unauthorized human gene therapy clinical trial was performed in 1980 [11, 23]. In this particular trial, a DNA encoding the herpes simplex virus thymidine kinase was electroporated in bone marrow cells from two β-thalassemia patients, and then reintroduced after subjecting the patients to local irradiation to favor engraftment. In fact, this approach had been previously shown to be ineffective in animal models, so no clear therapeutic benefits could be obtained from this clinical trial.

1.5 The Breakthroughs in Gene Therapy from the 1990s and 2000s

It is not surprising that the first choices for the application of gene therapy were those diseases that could be corrected by modification of hematopoietic cells. Gene mutations affecting hematopoiesis encompass a wide range of disorders, from hemophilia to immunodeficiency. Bone marrow cells can be effectively retrieved form human patients and easily manipulated in vitro. Bone marrow can be effectively depleted with chemotherapy and radiotherapy, and the hematopoietic compartment reconstituted by reintroduction of bone marrow cells. Thus, ex vivo gene transfer in bone marrow cells and their reintroduction can be effectively achieved. In addition, the potential benefits arising from gene therapy compared to the possible side effects were apparent, particularly in cases of severe immunodeficiency.

The first successful modification of hematopoietic stem cells and reconstitution of the hematopoietic compartment with genetically modified cells was reported in the early 1990s [24–27]. In this case, the neomycin resistance gene was inserted ex vivo using a γ-retrovirus in bone marrow from cancer patients. As part of their treatment, these patients had to undergo aggressive chemotherapy and bone marrow ablation. After gene marking, bone marrow cells were reintroduced in these patients, and expression of neomycin resistance was demonstrated in cells from the hematopoietic lineage. This was a critical experiment, which demonstrated that gene therapy could work in practical terms.

Thus, the first approved clinical trial was carried out in 1991 to correct severe combined immunodeficiency (ADA-SCID) [28–30]. In this case, the human adenosine deaminase gene was expressed using a γ-retrovirus. In this particular trial, $CD34^+$ cells (stem cell hematopoietic precursors) were isolated from peripheral blood followed by ex vivo retroviral transduction. Even though this clinical trial was successful from a safety point of view, it was unclear whether it brought true therapeutic benefits. Treated patients continued receiving exogenous ADA to control their disease.

Finally, after more than 40 years of extensive research, the major breakthrough in gene therapy was reported in 2000. X-SCID was successfully corrected in 11 children. Particularly, the treated type of severe immunodeficiency is caused by lack of expression of the common γ-chain interleukin receptor, which blocks differentiation of several cell types of the hematopoietic lineage, including T cells. In this case, the common interleukin receptor γ-chain was introduced in bone marrow using a retrovirus vector based on the MLV [31]. The same experimental approach and successful outcome was later reported by Adrian Thrasher's team in London [32].

1.6 Current Human Gene Therapy and Lentiviral Vectors

The initial high expectations from the first successful human gene therapy clinical trials using γ-retrovirus vectors were suddenly overshadowed by the development of leukemia in a significant number of children. In fact, the onset of leukemia was the direct result of the gene therapy itself. The retrovirus vector had integrated next to an oncogene, and its expression was upregulated in corrected lymphocytes, leading to their uncontrolled clonal expansion. This unexpected lethal side effect highlighted insertional mutagenesis as a major complication. Curiously, it has to be mentioned that in the early 1980s, the possibility of insertional mutagenesis had been raised as an important detrimental issue for the application of retrovirus vectors in gene therapy [11, 33, 34]. As a matter of fact, insertional mutagenesis had been used in vitro for the identification of growthpromoting and transforming genes [35]. Even so, clinical trials have continued and in 2006 X-linked chronic granulomatosis was successfully corrected in young adult patients, using γ-retrovirus vectors encoding gp91 phox, a subunit of the enzyme complex superoxide dismutase. Transduction of bone marrow cells with this retrovirus resulted in expression of pg91 in neutrophiles, which corrected their oxidative antibacterial capacities in the phagosome [36]. Moreover, clinical efficacy clearly correlated with clonal amplification of corrected cells, as the direct result of insertional activation of certain oncogenes with growth-promoting capacities [37]. Overall, these clinical trials highlighted an apparent necessity for corrected cells to have a selective advantage by increasing their proliferative capacity or promoting survival.

In 2006, it was demonstrated that efficacious gene therapy in humans was not restricted to modification of genetic mutations within the hematopoietic compartment. It was also successfully applied for the treatment of advanced melanoma in patients who were refractory to conventional antineoplastic

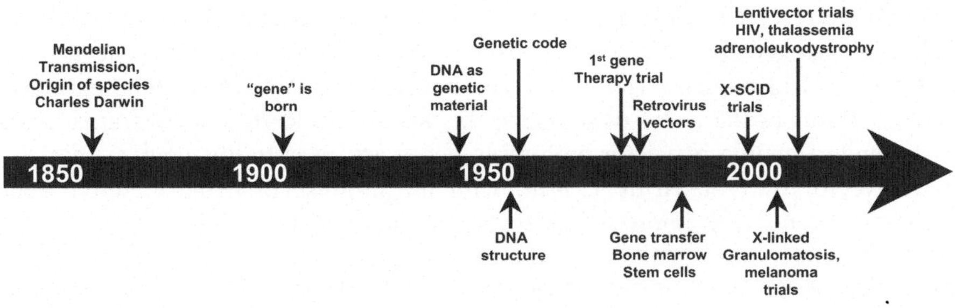

Fig. 1.1 Chronology of the main scientific steps leading to the development of gene therapy. This figure schematically and chronologically places the key scientific discoveries and groundbreaking therapeutic steps, which have led to the development of our current gene therapy protocols. Each significant discovery is indicated with arrows within an approximate time interval, spanning around the last 150 years

treatments. This trial consisted in the introduction of the α- and β-chains of a melanoma antigen (MART-1)-specific T cell receptor (TCR). This particular TCR was isolated and cloned from a patient that showed complete regression and long-term tumor-free survival after adoptive T cell transfer therapy. Peripheral blood lymphocyte preparations from 15 melanoma patients were transduced in vitro and reinfused. This therapy resulted in complete tumor regression and full recovery in two patients [38]. This clinical trial was truely a milestone. It clearly demonstrated that terminally differentiated T lymphocytes could be genetically modified to confer them effective anticancer activities.

It is in this context of relative success of retrovirus-based gene therapy that lentiviral vectors have come across. γ-retrovirus-based vectors present at least two major drawbacks. Firstly, they need the cells to be dividing to efficiently integrate their genome into the host cell chromosomes. This is not the case of lentivectors, which can transduce dividing and quiescent cells. Secondly, γ-retrovirus vectors are quite mutagenic, as shown in the human clinical trials. Even though lentivectors integrate in transcriptionally active sites, there is evidence that they are less mutagenic than their retrovirus counterparts. Therefore, lentivectors have started to be used in human clinical trials. To date, no adverse genotoxic effects have been reported from the first clinical trial in HIV-infected patients [39], and two clinical trials with lentivectors have resulted in full correction of β-thalassaemia [40] and X-linked adrenoleukodystrophy [41] (Fig. 1.1).

1.7 Concluding Remarks

The unprecedented rapid development of several scientific disciplines has enabled the isolation of genes and the manipulation of genetic information. Gene therapy came to life with high expectations that were progressively fading after several decades of research without noticeable therapeutic benefits. This situation has radically changed during the last decade, after the first successful gene therapy clinical trials in humans. However, the appearance of important genotoxic secondary effects as the result of the gene therapy itself clearly demonstrated that a better understanding of vector biology is necessary. It is in this context that lentiviral vectors have been developed, with the promise of being effective gene delivery systems with improved performance and biosafety.

Acknowledgments David Escors is funded by an Arthritis Research UK Career Development Fellowship (18433). Karine Breckpot is funded by the Fund for Scientific Research-Flandes.

References

1. Weiling F (1991) Historical study: Johann Gregor Mendel 1822–1884. Am J Med Genet 40(1):1–25 (discussion 26)
2. Falk R (1984) The gene in search of an identity. Hum Genet 68(3):195–204
3. Keeler CE (1947) Gene therapy. J Hered 38(10):294–298
4. Editorial (1976) Gene cloning: one milestone on a very long road. Lancet 1(7965):893
5. Avery OT, MacLeod CM, McCarty M (1979) Studies on the chemical nature of the substance inducing transformation of pneumococcal types. Inductions of transformation by a desoxyribonucleic acid fraction isolated from pneumococcus type III. J Exp Med 149(2):297–326
6. Watson JD, Crick FH (1953) Genetical implications of the structure of deoxyribonucleic acid. Nature 171(4361):964–967
7. Watson JD, Crick FH (1953) Molecular structure of nucleic acids; a structure for deoxyribose nucleic acid. Nature 171(4356):737–738
8. Grunberg-Manago M, Oritz PJ, Ochoa S (1955) Enzymatic synthesis of nucleic acidlike polynucleotides. Science 122(3176):907–910
9. Ochoa S (1963) Synthetic polynucleotides and the Genetic Code. Proceedings 5:37–64
10. Editorial (1981) Gene therapy: how ripe the time? Lancet 1(8213):196–197
11. Cline MJ (1985) Perspectives for gene therapy: inserting new genetic information into mammalian cells by physical techniques and viral vectors. Pharmacol Ther 29(1):69–92
12. Mulligan RC, Howard BH, Berg P (1979) Synthesis of rabbit beta-globin in cultured monkey kidney cells following infection with a SV40 beta-globin recombinant genome. Nature 277(5692):108–114
13. Hamer DH, Smith KD, Boyer SH, Leder P (1979) SV40 recombinants carrying rabbit beta-globin gene coding sequences. Cell 17(3):725–735
14. Mantei N, Boll W, Weissmann C (1979) Rabbit beta-globin mRNA production in mouse L cells transformed with cloned rabbit beta-globin chromosomal DNA. Nature 281(5726):40–46
15. Nagata S, Taira H, Hall A, Johnsrud L, Streuli M, Ecsodi J, Boll W, Cantell K, Weissmann C (1980) Synthesis in *E. coli* of a polypeptide with human leukocyte interferon activity. Nature 284(5754):316–320
16. Friedmann T (1976) The future for gene therapy–a reevaluation. Ann N Y Acad Sci 265:141–152
17. Friedmann T, Roblin R (1972) Gene therapy for human genetic disease? Science 175(25):949–955
18. Terheggen HG, Lowenthal A, Lavinha F, Colombo JP, Rogers S (1975) Unsuccessful trial of gene replacement in arginase deficiency. Zeitschrift fur Kinderheilkunde 119(1):1–3

19. Neville R (1976) Gene therapy and the ethics of genetic therapeutics. Ann N Y Acad Sci 265:153–169
20. Mann R, Mulligan RC, Baltimore D (1983) Construction of a retrovirus packaging mutant and its use to produce helper-free defective retrovirus. Cell 33(1):153–159
21. Pear WS, Nolan GP, Scott ML, Baltimore D (1993) Production of high-titer helper-free retroviruses by transient transfection. Proc Natl Acad Sci U S A 90(18):8392–8396
22. Williamson B (1982) Gene therapy. Nature 298(5873):416–418
23. Mercola KE, Cline MJ (1980) Sounding boards. The potentials of inserting new genetic information. N Engl j med 303(22):1297–1300
24. Brenner MK, Rill DR, Holladay MS, Heslop HE, Moen RC, Buschle M, Krance RA, Santana VM, Anderson WF, Ihle JN (1993) Gene marking to determine whether autologous marrow infusion restores long-term haemopoiesis in cancer patients. Lancet 342(8880):1134–1137
25. Deisseroth AB, Zu Z, Claxton D, Hanania EG, Fu S, Ellerson D, Goldberg L, Thomas M, Janicek K, Anderson WF et al (1994) Genetic marking shows that Ph + cells present in autologous transplants of chronic myelogenous leukemia (CML) contribute to relapse after autologous bone marrow in CML. Blood 83(10):3068–3076
26. Dunbar CE, Cottler-Fox M, O'Shaughnessy JA, Doren S, Carter C, Berenson R, Brown S, Moen RC, Greenblatt J, Stewart FM et al (1995) Retrovirally marked CD34-enriched peripheral blood and bone marrow cells contribute to long-term engraftment after autologous transplantation. Blood 85(11):3048–3057
27. Rill DR, Santana VM, Roberts WM, Nilson T, Bowman LC, Krance RA, Heslop HE, Moen RC, Ihle JN, Brenner MK (1994) Direct demonstration that autologous bone marrow transplantation for solid tumors can return a multiplicity of tumorigenic cells. Blood 84(2):380–383
28. Anderson WF, Blaese RM, Culver K (1990) The ADA human gene therapy clinical protocol: points to consider response with clinical protocol, July 6, 1990. Hum Gene Ther 1(3):331–362
29. Blaese RM, Culver KW, Chang L, Anderson WF, Mullen C, Nienhuis A, Carter C, Dunbar C, Leitman S, Berger M et al (1993) Treatment of severe combined immunodeficiency disease (SCID) due to adenosine deaminase deficiency with CD34 + selected autologous peripheral blood cells transduced with a human ADA gene. Amendment to clinical research project, Project 90-C-195, January 10, 1992. Hum Gene Ther 4(4):521–527
30. Levine F, Friedmann T (1991) Gene therapy techniques. Curr Opin Biotechnol 2(6):840–844
31. Cavazzana-Calvo M, Hacein-Bey S, de Saint Basile G, Gross F, Yvon E, Nusbaum P, Selz F, Hue C, Certain S, Casanova JL, Bousso P, Deist FL, Fischer A (2000) Gene therapy of human severe combined immunodeficiency (SCID)-X1 disease. Science 288(5466):669–672
32. Gaspar HB, Parsley KL, Howe S, King D, Gilmour KC, Sinclair J, Brouns G, Schmidt M, Von Kalle C, Barington T, Jakobsen MA, Christensen HO, Al Ghonaium A, White HN, Smith JL, Levinsky RJ, Ali RR, Kinnon C, Thrasher AJ (2004) Gene therapy of X-linked severe combined immunodeficiency by use of a pseudotyped gammaretroviral vector. Lancet 364(9452):2181–2187
33. Howe SJ, Mansour MR, Schwarzwaelder K, Bartholomae C, Hubank M, Kempski H, Brugman MH, Pike-Overzet K, Chatters SJ, de Ridder D, Gilmour KC, Adams S, Thornhill SI, Parsley KL, Staal FJ, Gale RE, Linch DC, Bayford J, Brown L, Quaye M, Kinnon C, Ancliff P, Webb DK, Schmidt M, von Kalle C, Gaspar HB, Thrasher AJ (2008) Insertional mutagenesis combined with acquired somatic mutations causes leukemogenesis following gene therapy of SCID-X1 patients. J Clin Invest 118(9):3143–3150
34. Hacein-Bey-Abina S, Von Kalle C, Schmidt M, McCormack MP, Wulffraat N, Leboulch P, Lim A, Osborne CS, Pawliuk R, Morillon E, Sorensen R, Forster A, Fraser P, Cohen JI, De Saint Basile G, Alexander I, Wintergerst U, Frebourg T, Aurias A, Stoppa-Lyonnet D, Romana S, Radford-Weiss I, Gross F, Valensi F, Delabesse E, Macintyre E, Sigaux F, Soulier J, Leiva LE, Wissler M, Prinz C, Rabbitts TH, Le Deist F, Fischer A, Cavazzana-Calvo M (2003) LMO2-associated clonal T cell proliferation in two patients after gene therapy for SCID-X1. Science 302(5644):415–419

35. Stocking C, Loliger C, Kawai M, Suciu S, Gough N, Ostertag W (1988) Identification of genes involved in growth autonomy of hematopoietic cells by analysis of factor-independent mutants. Cell 53(6):869–879
36. Moreno-Carranza B, Gentsch M, Stein S, Schambach A, Santilli G, Rudolf E, Ryser MF, Haria S, Thrasher AJ, Baum C, Brenner S, Grez M (2009) Transgene optimization significantly improves SIN vector titers, gp91phox expression and reconstitution of superoxide production in X-CGD cells. Gene Ther 16(1):111–118
37. Ryser MF, Roesler J, Gentsch M, Brenner S (2007) Gene therapy for chronic granulomatous disease. Expert Opin Biol Ther 7(12):1799–1809
38. Morgan RA, Dudley ME, Wunderlich JR, Hughes MS, Yang JC, Sherry RM, Royal RE, Topalian SL, Kammula US, Restifo NP, Zheng Z, Nahvi A, de Vries CR, Rogers-Freezer LJ, Mavroukakis SA, Rosenberg SA (2006) Cancer regression in patients after transfer of genetically engineered lymphocytes. Science 314(5796):126–129
39. Wang GP, Levine BL, Binder GK, Berry CC, Malani N, McGarrity G, Tebas P, June CH, Bushman FD (2009) Analysis of lentiviral vector integration in HIV + study subjects receiving autologous infusions of gene modified CD4 + T cells. Mol Ther 17(5):844–850
40. Cavazzana-Calvo M, Payen E, Negre O, Wang G, Hehir K, Fusil F, Down J, Denaro M, Brady T, Westerman K, Cavallesco R, Gillet-Legrand B, Caccavelli L, Sgarra R, Maouche-Chretien L, Bernaudin F, Girot R, Dorazio R, Mulder G.J, Polack A, Bank A, Soulier J, Larghero J, Kabbara N, Dalle B, Gourmel B, Socie G, Chretien S, Cartier N, Aubourg P, Fischer A, Cornetta K, Galacteros F, Beuzard Y, Gluckman E, Bushman F, Hacein-Bey-Abina S, Leboulch P (2010) Transfusion independence and HMGA2 activation after gene therapy of human beta-thalassaemia; 1476-4687 (Electronic) 0028-0836 (Linking), Sept 16 2010, pp 318–322
41. Cartier N, Hacein-Bey-Abina S, Bartholomae CC, Veres G, Schmidt M, Kutschera I, Vidaud M, Abel U, Dal-Cortivo L, Caccavelli L, Mahlaoui N, Kiermer V, Mittelstaedt D, Bellesme C, Lahlou N, Lefrere F, Blanche S, Audit M, Payen E, Leboulch P, l'Homme B, Bougneres P, Von Kalle C, Fischer A, Cavazzana-Calvo M, Aubourg P (2009) Hematopoietic stem cell gene therapy with a lentiviral vector in X-linked adrenoleukodystrophy. Science 326(5954):818–823

Chapter 2
Development of Retroviral and Lentiviral Vectors

David Escors, Grazyna Kochan, Holly Stephenson and Karine Breckpot

Abstract Gene vectors based on human immunodeficiency virus 1 (HIV-1) are becoming popularly used as gene carriers. HIV-1 lentivectors have recently been used in two gene therapy clinical trials for the correction of β-thalassaemia and X-linked adrenoleukodystrophy. The process of transforming a deadly human pathogen such as HIV into a successful therapeutic tool would not be possible without thorough scientific investigation into the development of γ-retrovirus vectors. In this chapter, we briefly recapitulate the major scientific steps that have led to the development of γ-retrovirus and lentivirus vectors.

2.1 Retrovirus Biology

2.1.1 Brief Introduction to Retroviruses

The hallmark characteristic of retroviruses is their capacity to retrotranscribe their RNA genome into a cDNA copy, which is stably integrated into the host cell chromosome. This makes them ideal gene carriers into target cells, a key requirement for

D. Escors (✉)
University College London, Rayne Building, 5 University Street, London WC1E 6JF, UK
e-mail: d.escors@ucl.ac.uk

G. Kochan
Oxford Structural Genomics Consortium, University of Oxford,
Old Road Campus Research Building, Roosevelt Drive, Headington, Oxford OX3 7DQ, UK

H. Stephenson
Institute of Child Health, University College London, Great Ormond Street,
London WC1N 3JH, UK

K. Breckpot
Vrije Universiteit Brussels, Brussels, Belgium

D. Escors et al., *Lentiviral Vectors and Gene Therapy*,
SpringerBriefs in Biochemistry and Molecular Biology,
DOI: 10.1007/978-3-0348-0402-8_2,

successful gene therapy. Vectors based on γ-retroviruses were the first used in human gene therapy. However, their significant genotoxicity has led researchers back to the drawing board for major improvements in design, performance, and biosafety. Nevertheless, it is not by chance that retrovirus vectors were amongst the first to be engineered and applied in human therapy. Their genome organization is fairly simple and their life cycle is well known due to historical reasons. Retroviruses were thoroughly studied very early as some strains caused cancer in many vertebrate species. As a matter of fact, the first oncogenes were found in some retroviral species which were consequently named oncoretroviruses [1–6]. Cellular counterparts were soon discovered after the identification and isolation of retroviral oncogenes [7–12]. Without any doubt, research in retrovirus biology was directly responsible for the discovery of critical intracellular signaling molecules and transcription factors such as Src, Ras, Raf, c-Rel (NF-κB), and c-myc.

In the early 1980s, a pandemic causing severe immunodeficiency in humans spread worldwide. This disease was called acquired immunodeficiency syndrome (AIDS) and raised unprecedented concern. The infectious agent was identified as a retrovirus related to simian immunodeficiency virus (SIV), a member of the lentivirus genus. This virus was originally termed human T lymphotropic virus (HTLV)-3, and later human immunodeficiency virus 1 (HIV-1) [13–17]. Prior to the discovery of the HIV retrovirus, only HTLV-1 and 2 were known to cause severe human disease. Even though antiretroviral therapy can effectively control HIV infection, AIDS is still a medical burden of enormous proportions. No cure has yet been found, and only one case of complete recovery has ever been reported in a patient undergoing bone marrow transplant from a resistant donor [18]. A combination of extensive research in HIV biology culminated in the development of HIV-1 lentiviral vectors. Why was HIV-1 chosen for the development of gene transfer vectors? HIV possesses unique qualities, which makes it ideal as a gene transfer vector once the pathogenic genes are removed. Consequently, it has been successfully used in at least two human clinical trials [19, 20].

2.1.2 The Retrovirus Virion

Both γ-retroviruses and lentiviruses belong to the *Retroviridae* family, which encompasses a group of spherical enveloped viruses with a diameter between 80 and 120 nm (Fig. 2.1a) [21]. Retroviruses contain a diploid genome made of two identical molecules of single positive-stranded RNA. The genome is present within an internal core made of several structural and enzymatic proteins: nucleocapsid (NC), capsid (CA), reverse transcriptase (RT), integrase (IN), and protease (PR) (Fig. 2.1a). In addition, there is an outer protein layer made of matrix (MA) protein, which interacts with the internal core and with envelope glycoprotein (ENV) in the virion lipid envelope. ENV binds to the cellular receptor and mediates fusion between the virion envelope and the cellular membrane. ENV is post-translationally processed into two subunits; a transmembrane subunit (TM) which is anchored to the virion lipid envelope, and a globular subunit (SU), which binds to the cellular receptor. The SU subunit remains noncovalently attached to TM (Fig. 2.1a).

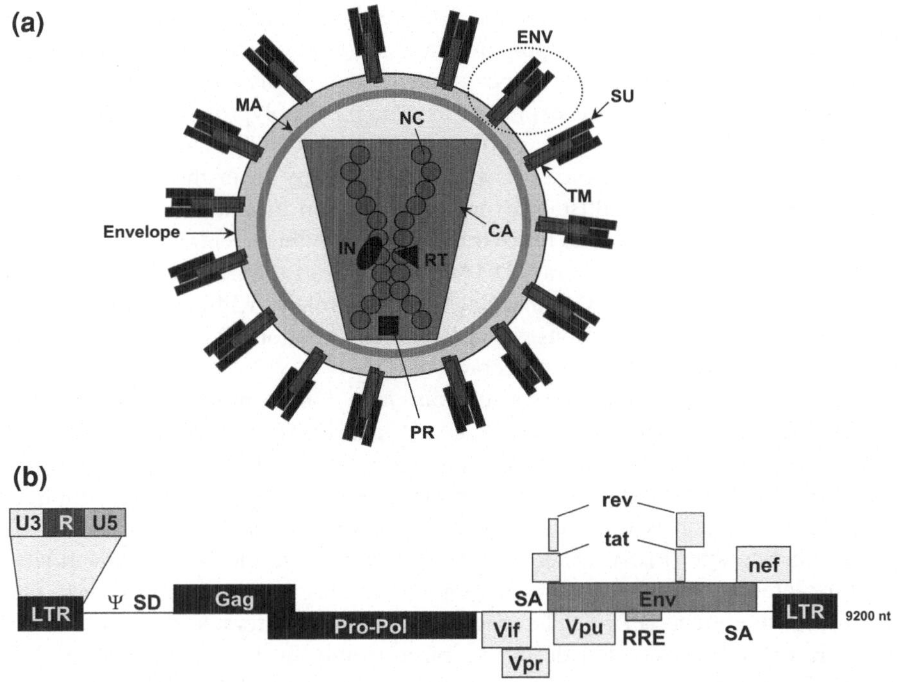

Fig. 2.1 Retrovirus virion structure and HIV-1 genome organization. **a** Retrovirus virion structure represented as a sphere containing the internal core (*conical gray* structure, as organized in HIV-1 [33–35]) made of the NC, CA, IN, RT, and PR. The NC protein binds to two strands of single-stranded positive-sense genome RNA molecules. The internal core is contained within a layer of matrix protein (MA) underneath the virus envelope. The envelope glycoprotein (ENV, within a *dotted ellipse*) is made of the spherical subunit (SU) which is attached to the transmembrane domain (TM). TM is inserted in the virus envelope. **b** The HIV-1 provirus genome organization is shown as a model for a complex retrovirus. From the left to the right: LTR indicates the long terminal repeats which flank both ends of the provirus. On the left LTR, its functional organization is highlighted (U3, R, and U5); Ψ (psi), indicates the HIV-1 packaging signal; SD, splicing donor sequence, involved in transcription of the Env mRNA and some of the mRNAs encoding accessory genes; Gag, Gag polyprotein encoding MA-CA-NC. During its transcription, an RNA frameshifting takes place after a pseudoknot formation between Gag and Pro-Pol leading to an mRNA encoding Gag–Pro-Pol; SA, splicing acceptor sequence; Env, retrovirus envelope glycoprotein gene. The rest of the genes (vif, vpr, vpu, rev, tat, and nev) are specific for HIV-1 and represent accessory and virulence genes, not present in simple retroviruses; RRE represents rev response element, a region to which rev binds, inhibiting splicing, and regulating the rate of Gag–Pol/Env, accessory genes RNA transcription

2.1.3 The Organization of the Retrovirus Genome

According to their genome organization, retroviruses can be classified into two main groups; simple and complex retroviruses. In either case, their genome organization is fairly similar overall. What separates complex from simple retroviruses is the presence of a variable number of accessory and regulatory genes. The genome of simple retroviruses is organized from the 5′ to the 3′ end with Gag, Pol, and Env genes (Fig. 2.1b) [22]. Gag encodes a polyprotein made of the main structural retrovirus proteins, MA, CA, and NC. Pol encodes enzymatic proteins associated to the RNA genome within the virion. It is expressed as a Gag–Pol polyprotein, synthesized by ribosomal frameshifting during Gag mRNA translation [23]. Pol encodes the RT, IN, and PR proteins. RT synthesizes a single cDNA copy from the two strands of genomic RNA using a multistep reverse transcription mechanism [24, 25]. This reaction takes place within the retrovirus core soon after its entrance into the cell cytoplasm. IN mediates the viral cDNA integration in the host cell chromosome (provirus DNA). PR cleaves Gag and Gag–Pol polyproteins within the released viral particle leading to fully infectious virions (virion maturation). As mentioned earlier, Env encodes the virus envelope glycoprotein ENV.

The integrated provirus genome contains two long terminal repeats (LTRs) subdivided in three functional regions, U3, R, and U5 (Fig. 2.1b); briefly, the U3 region is in fact the functional HIV-1 promoter, and contains the transcriptional enhancers and TATA box. The R region marks the starting point of transcription, while the U5 region is involved in reverse transcription and contains the tRNA primer-binding site (PBS). Other important elements are the IN attachment sites used for integration into the host cell genome, the packaging signal (Ψ, psi), and the polypurine tract (PPT). The packaging signal confers specificity for the genomic RNA encapsidation during virion assembly in the cell cytoplasm [26]. The PPT element is the site of initiation of positive-strand DNA synthesis during reverse transcription [27, 28].

The complex retrovirus genome, such as that of HIV and human T cell lymphotropic viruses, shares the same basic organization. However, it contains an additional set of accessory genes involved in regulation of transcription, RNA transport, gene expression, and assembly [29]. Examples of these are HTLV rex and tax [30–32], or HIV-1 tat, rev, vif, vpr, vpu and nef (Fig. 2.1b).

2.1.4 The Retrovirus Life Cycle

A scheme of the retrovirus life cycle is depicted in Fig. 2.2. The life cycle starts with virion binding to the cellular receptor through its envelope glycoprotein. The type of receptor to which ENV binds determines the virus cell and tissue tropism. Many retroviral cellular receptors are very well characterized. For example, the murine cationic amino acid transporter is the receptor for mouse leukemia virus (MLV) ecotropic envelope [36]; sodium/phosphate symporters, for the MLV amphotropic, and gibbon ape leukemia virus (GALV) envelope glycoproteins; and the T lymphocyte receptor CD4 and coreceptors CXCR4 and CCR5 for HIV-1 gp120 [37–39], to name just a few examples.

ENV binding to its receptor results in a conformational change that exposes the fusion peptide, present in the TM subunit [40, 41]. Subsequently, fusion between

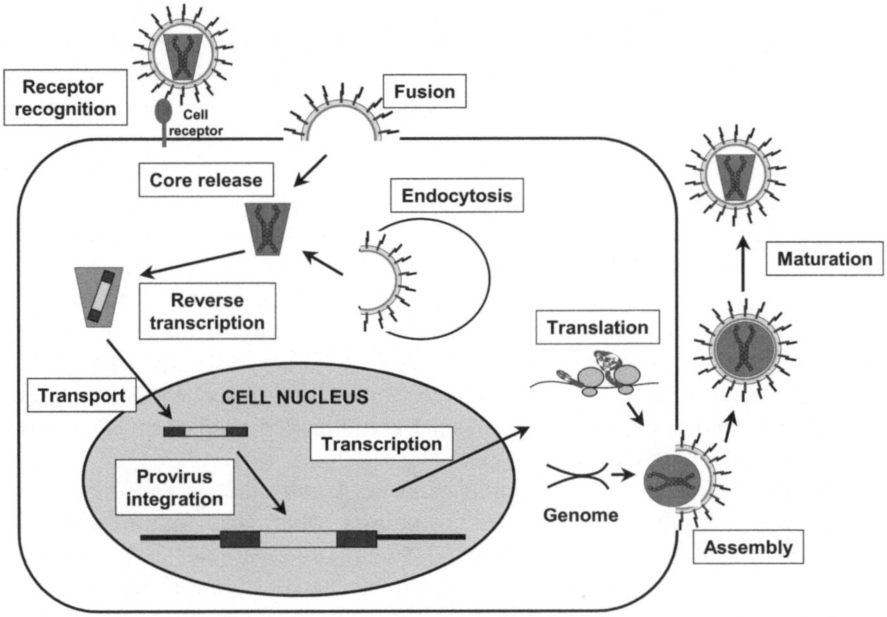

Fig. 2.2 General retrovirus life cycle. A simplified scheme of the retroviral life cycle is shown, divided into the major steps which have been indicated by closed text boxes. The retrovirus virion binds specifically to the target cell after receptor recognition (*upper part* of the scheme), followed by either direct fusion (Fusion box) with the cell membrane, or endocytosis followed by membrane fusion (Endocytosis box). Either way, the retrovirus core is released into the cell cytoplasm (Core release), and the RNA genome undergoes reverse transcription generating a cDNA copy (Reverse transcription box. The RNA genome is shown in red and the cDNA counterpart in blue). In the case of HIV, the core is actively transported into the cell nucleus (Transport), while in the case of oncoretroviruses, the nucleus membrane has to disappear during cell mitosis for provirus integration. In the nucleus, the cDNA genome is integrated and stays in the target cell chromosome as a provirus (Provirus integration). Once integrated, transcription takes place leading to full RNA genome copies, or mRNAs which after translation give rise to structural and nonstructural proteins (Transcription and Translation). Gag and Gag–Pol products assemble at the cellular membrane and encapsidate the virus genome, leading to virion release and maturation after PR-mediated cleavage of structural and enzymatic virion proteins

the virion and cellular membrane takes place, and the retrovirus core is released into the cytoplasm. The RNA genome is reverse transcribed within the viral core. From the two RNA genome molecules, one double-stranded cDNA copy is produced [25, 28]. This cDNA copy from within the viral core is transported and integrated into the cell chromosome. Here lies one of the main differences between simple retroviruses and lentiviruses. Cell division and concomitant disappearance of the nuclear membrane are absolutely required for the integration of γ-retrovirus genomes [42]. In contrast, the lentivirus preintegration complex is actively transported into the cell nucleus, without the need for cell mitosis [43]. As a consequence, virus vectors based on γ-retroviruses only transduce dividing cells, while lentivectors transduce dividing and nondividing (quiescent) cells.

Once in the nucleus, the viral cDNA integrates into the host cell DNA through the enzymatic activities of IN. The term provirus describes the integrated retrovirus, from which viral genes are transcribed and spliced utilizing the cellular transcriptional and posttranscriptional machinery. Therefore, gene expression can take place during the life of the infected cell and its progeny. Interestingly, if integration occurs in the germline cells, infectious retroviruses can be hereditary and transmitted as mendelian traits. This has taken place multiple times through evolution and accounts for the large quantity of retrovirus-like elements even in the human genome [44]. Interestingly, some of these have been positively selected in evolution, and now play physiological roles, such as syncytin, an envelope protein from an endogenous human defective retrovirus involved in placenta morphogenesis [45]. Another classical example is the tissue-specific promoter for the human amylase gene, which is derived from retroviral sequences [46]. Actually, the integration of retroviruses into the germline is actively occurring nowadays in other mammalian species [47].

The transcribed full-length genomic viral RNA and mRNAs encoding all the viral proteins are transported to the cell cytoplasm, where they are translated. The Gag and Gag–Pol polyproteins assemble into virus-like particles and specifically package the unspliced genomic RNA containing the packaging signal (Fig. 2.2). Virion budding then takes place at the cell membrane where the particle incorporates its lipid envelope containing ENV. Interestingly, retroviruses passively can acquire a wide range of different viral glycoproteins if expressed in the infected cells, a process called pseudotyping [48–52]. Finally, virion maturation occurs after budding from the cell by the activity of the viral protease, which proteolyses Gag and Gag–Pol polyproteins leading to MA, CA, NC, and enzymatic (RT, IN) proteins (Fig. 2.2, maturation) [53, 54].

2.2 Vectors Based on γ-retroviruses

2.2.1 Development of γ-retroviral Vectors

As discussed earlier, retrovirus research gained a tremendous momentum during the early 1980s due to their participation in cellular transformation and oncogenesis. Some retroviruses can transform cells by either integrating oncogenes directly into the host cell chromosomes or by insertional mutagenesis after viral genome integration [55]. Interestingly, their capacity to integrate their RNA genome as a cDNA version into the host cell chromosomes was exploited for the efficient introduction of exogenous DNA into mammalian cells. Consequently, the engineering of gene vectors derived from Moloney Mouse Leukaemia Virus (MLV) was carried out in the early 1980s [56]. Briefly, to generate a virus vector suitable for gene therapy applications, it is fundamental to separate the transfer vector (containing the promoter and gene of interest) from the structural and enzymatic genes required for virus propagation. This will ensure that nonreplicating infectious viruses are generated. Thus, the virus vector (the transfer vector) essentially contains a promoter controlling transgene transcription and all the *cis*-acting sequences necessary for its replication/retrotranscription and packaging into virus-like particles. These elements include the LTRs, packaging signal and sequences involved in reverse transcription, and integration. Early packaging systems were fairly unsophisticated. The first one consisted of murine NHI 3T3 cell clones transfected with a molecular clone of a Moloney MLV genome lacking the packaging signal [56]. These cells did not produce detectable infectious virions during the early cell passages, but could effectively package a defective Moloney sarcoma virus, which did not encode intact retroviral proteins [56]. This first system demonstrated that replicase and packaging activities could be provided *in trans* by a packaging cell line. Nevertheless, replication-competent virus was eventually rescued from these packaging cells possibly after recovery of packaging sequences from endogenous retrovirus-like elements. Further refinement of this early retrovirus vector system was concentrated in the production of helper-free retrovirus vector preparations [57]. This system was based on transient transfection of three plasmids in a highly transfectable 293T cell clone; a first plasmid encoding Gag–pol but containing a mutation in ENV, a second plasmid containing a mutation in the Gag–pol but with the ecotropic ENV gene intact, and a third plasmid, the transfer vector, expressing selected markers from the viral LTR. Thus, the first two provided just the packaging functions (Gag–pol, env) *in trans* [57]. Nevertheless, there was still a relatively high chance of recombination that could lead to replication-competent viruses. A major improvement on this system came when Gag–Pol, Env, and the transfer vectors were expressed by transient transfection under the control of the cytomegalovirus promoter (CMV). In this way, most of the MLV sequences including the retrovirus LTRs were removed from the packaging constructs, reducing the likelihood of recombination leading to replication-competent viruses [58]. This system provides the basis for the development of packaging cells currently used. Further improvements to the packaging systems and vector design were implemented to increase the efficiency and biosafety. Equivalent improvements were applied for the generation of lentiviral vectors and will be discussed in some detail below.

2.2.2 Advantages and Limitations of γ-retroviral Vectors

There are several advantages of the use of γ-retroviral vectors. Firstly, they stably integrate their genome into the cell chromosome, leading to long-term transgene expression. Secondly, the transfer vector is nonreplicative and does not encode viral proteins. This lack of retroviral proteins decreases the antivector immunogenicity, which is a major issue for other viral systems such as adenovirus and vaccinia virus vectors. In any case, antiretrovirus vector immunity is not completely avoided, and neutralizing antibodies are generated against the envelope protein after consecutive uses. To avoid this retrovirus vectors can be easily pseudotyped with a variety of envelopes.

Even though γ-retrovirus vectors are largely a success both in animal models of disease and in human gene therapy (apart from their serious genotoxic effects), they have some critical limitations. Retrovirus particles are rather unstable, and the use of some stabilizing agents commonly used in cell cultures complicate their application in vivo [59]. γ-retrovirus vectors do not achieve high viral titers [60], usually on the range of 10^5 particles per ml, unlike adenovirus or poxvirus-based vectors, reaching titers up to 10^{10}–10^{12} [61, 62]. These relatively low titers require production of large volumes of supernatants from producer cells, followed by vector concentration usually by ultracentrifugation [48]. However, the two major disadvantages for the use of γ-retrovirus vectors are (1) their inability to transduce nondividing cells [42], and (2) insertional mutagenesis [63, 64]. γ-retrovirus vectors have been particularly successful in human gene therapy clinical trials, at least from the therapeutic point of view [65–67]. In these cases, highly proliferating hematopoietic stem cells were successfully targeted for genetic correction. However, γ-retrovirus vectors are rather unsuitable for the genetic modification of highly differentiated, undividing cells such as neurons or muscle. This highlights the greatest limitation on the use of retroviral vectors is their transforming capacity. Retroviral vectors can induce insertional mutagenesis in the target cell, by upregulating the expression of protooncogenes or the inactivation of antioncogenes. This is a well-known phenomenon, used to identify growth factors and protooncogenes [55]. In fact, insertional mutagenesis has caused serious complications in human gene therapy. Integration of the therapeutic retrovirus close to protooncogenes in hematopoietic stem cells has caused leukemia in a significant number of treated children for the correction of X-SCID [64, 67, 68].

2.3 Vectors Based on Lentiviruses

2.3.1 Development of Lentivectors

As discussed above, lentiviruses are complex retroviruses and encode a range of regulatory proteins involved in virus replication, assembly, and virulence. The most widely used lentivectors are those based on the HIV-1 genome. While quite a significant scientific effort was invested in overcoming the shortcomings of γ-retrovirus vectors, especially the inefficient transduction of quiescent cells, lentiviruses exhibited a remarkable capacity for transduction of highly differentiated, quiescent cells [43, 69, 70]. Lentivectors retained the capacity of efficient transduction of quiescent cells even in vivo, such as neurons [71–73]. Lentivector transduction of nondividing cells requires the active transport of the preintegration complex into the cell nucleus. This process is driven by the integrase protein [43], MA [43, 69], Vpr [74] and the central polypurine tract (cPPT) sequence [75]. However, progression through the cell cycle seems to be required for some cells [76].

The basic principles for the lentivector development are the same as those for γ-retrovirus vectors. Briefly, lentivectors are usually generated by a three-plasmid cotransfection system in 293T cells [71, 72] (Fig. 2.3); a packaging plasmid providing the RT and structural proteins, an envelope plasmid, encoding a viral glycoprotein for the pseudotyping of virus particles, and the transfer plasmid, containing all the *cis*-acting sequences for replication/transcription and packaging. The later generation systems are based on either a four-plasmid transfection, providing rev and tat in an additional plasmid [77], or by continuous stable production from cell lines [78].

2.3.2 Lentivector Generation Systems

Lentivector production systems have been refined over time to improve their performance and biosafety (Fig. 2.3). This is particularly important for HIV-1 lentivectors, since it is a human pathogen to which no cure is currently available. In the first generation of vectors, the packaging plasmid provided Gag–Pol and the accessory genes vif, vpu, vpr, nef, rev, and tat. Only the ENV gene was removed [72]. These genes are involved in HIV-1 virulence and first generation lentivectors posed a significant biosafety risk. Fortunately, most accessory genes could be removed in the second generation system (Fig. 2.3a) without affecting lentivector performance [79, 80]. Even if replication-competent viruses arise from multiple recombination events, these viruses would be devoid of all accessory genes. The third generation system (Fig. 2.3b) further improved biosafety, by replacing the U3 sequence in the transfer plasmid LTR by a nonHIV strong constitutive promoter [81]. Thus, tat could be eliminated as transcription did not depend anymore from the HIV U3, which requires tat [82, 83]. In addition to that, rev could be provided *in trans* in a separate plasmid, further reducing the possibility of recombination leading to replication-competent lentiviruses.

2.3.3 Improvements and Modifications of Lentiviral Vectors

As discussed earlier, the transfer vector contains all sufficient and necessary *cis*-acting sequences for its reverse transcription and packaging into lentivector particles. However, for translation into a viable clinical vector, titers of around 10^6 transducing particles per ml have to be achieved. Therefore, a substantial effort has been invested in optimizing lentivector production and transduction efficiency (Fig. 2.3c). A critical limiting step is nuclear import of the genome cDNA, mediated by nuclear transport signals present in IN, CA, and Vpr proteins. In addition, the central PPT and the central termination sequence (CTS) also enhance import of the preintegration complex [43, 69, 75]. These sequences form a *cis*-acting DNA flap that increases titers and allows significant transduction of CD34+ human hematopoietic cells (Fig. 2.3c) [84].

Lentivector transfer molecules can accommodate heterologous sequences that stabilize/improve vector RNA processing, and transport out of the nucleus. One of such sequences is the posttranscriptional regulatory element present in S transcripts

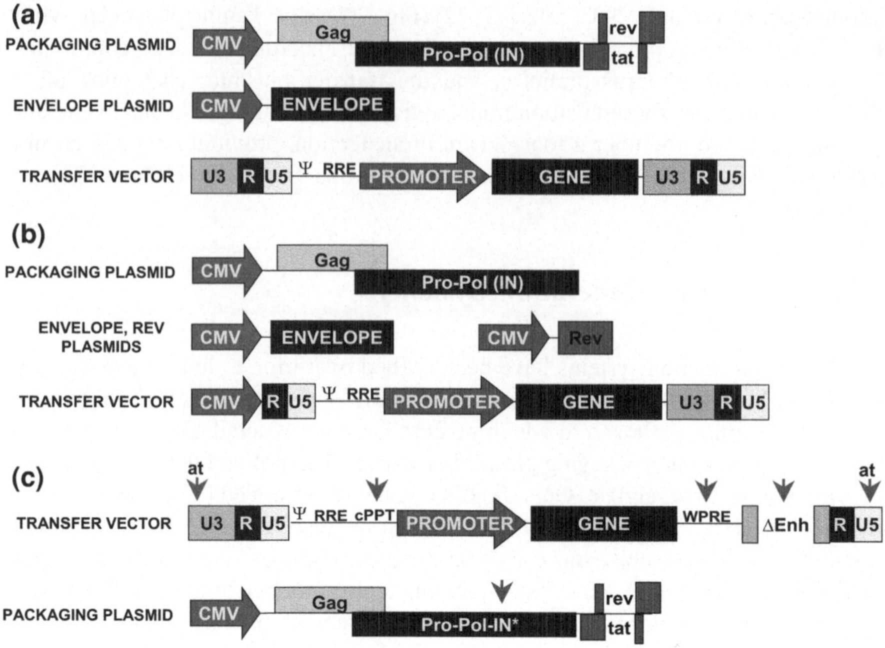

Fig. 2.3 Most commonly used lentivector systems and modifications to improve lentivector performance and biosafety. **a** The second generation lentivector system composes cotransfection of three different plasmids, as depicted in the figure; a packaging plasmid, expressing Gag–Pol, rev, and tat under the control of a strong nonretroviral promoter (CMV in this case); an envelope plasmid, expressing a wide range of envelope glycoproteins from a nonretroviral promoter such as CMV; the transfer vector plasmid, containing the two LTRs, packaging signal (Ψ), rev response element (RRE), and internal promoter of choice leading to expression of the gene of interest. This system is dependent on tat expression, so that efficient transcription from HIV-1 U3 can take place in lentivector producer cells. **b** The third generation lentivector system differs from the second in the absence of rev and tat from the packaging construct, and the replacement of the 5′ HIV U3 by a strong constitutive promoter such as CMV in this case. This system is tat independent because, transcription of the RNA transfer genome takes place constitutively from CMV. In the fourth generation, rev is provided in an additional separate plasmid. **c** Further improvements in the lentivector system are depicted in this figure, and indicated with red arrows. Firstly, the transfer vector can be modified to include a central polypurine tract (cPPT) and the woodchuck posttranscriptional regulatory element (WPRE), which increase lentivector titers and gene expression. In addition, the enhancers from the 3′ HIV U3 can also be removed (ΔEnh), leading to a self-inactivating lentivector. Taking advantage of the fact that the 5′ LTR from the integrated provirus comes from the 3′ LTR in the genomic RNA, the enhancer-deleted U3 region is copied in the integrated provirus, and thus, the HIV U3 promoter is absent in transduced cells. Secondly, mutations in the integrase attachment sites (at, indicated by *red arrows*) prevent the integration of the transfer vector, remaining in the cell nucleus as an episome (nonintegrated lentivector). Finally, the packaging plasmid can be modified to introduce mutations in IN which prevent insertion of the transfer vector resulting in nonintegrating lentivectors

from hepatitits B virus (Fig. 2.3c). This element facilitates RNA transport, inhibits splicing, increases protein expression, and replaces some of rev functions [85]. Incorporation of the woodchuck hepatitis virus posttranscriptional element (WPRE) in the transfer vector increased transgene expression between 3 and 8 fold [86]. However, there has been some concern that a gene product encoded by the WPRE element, the X protein, could increase lentivector genotoxicity. To overcome this problem, a deleted version of the WPRE lacking X gene and its promoter sequence has been engineered, which retained its capacity to increase lentivector titers [87].

Other changes in lentivector systems have been directed to increasing biosafety (Fig. 2.3c), especially after the first clinical application of γ-retroviruses for correction of X-linked SCID and chronic granulomatous disease (CGD). Even though X-SCID was corrected in treated children, a significant number of children developed leukemia which was linked to upregulation of oncogene expression by insertional mutagenesis [64, 88–90]. One way to prevent this phenomenon is to eliminate proviral LTRs, because they contain strong transcriptional viral enhancers that can upregulate protooncogenes. To achieve this, self-inactivating transfer vectors have been constructed by deleting transcription factor binding sites and the TATA box in the 3′ U3 region [91–93]. After integration, the provirus duplicates the U3-deleted 3′LTR in its 5′ end, resulting in nonfunctional deleted proviral LTRs. This approach minimizes the appearance of replication-competent viruses by recombination, and avoids nearby oncogene transactivation after provirus integration. Other strategies have attempted to completely abrogate integration by developing nonintegrating lentivectors (NILVs). To produce NILVs the packaging plasmid contains inactivating mutations in selected regions of the IN coding region, or the mutations are introduced in the ends of the transfer vector itself binding IN [94, 95]. Thus, the transfer vector does not integrate and remains in the nucleus as an episome. However, although long-term transgene expression is achieved in postmitotic cells and tissues, it does not in dividing cells because the vector is diluted between proliferating cells. This strategy has shown to be effective for gene therapy in retina, muscle, and brain [94, 96–98]. Additionally, NILVs could make ideal vaccines. They elicit strong immune responses and their inability to induce long-term transgene expression in dividing cells is an advantage, as this can be detrimental once the immune response is terminated [99]. Vaccination with NILVs elicits antigen-specific T cell responses and is effective for the treatment of lymphoma in mouse experimental models [95, 100, 101].

2.3.4 Advantages and Limitations of Lentivectors

Lentivectors share many of the advantages with γ-retrovirus counterparts, including stable integration, long-term transgene expression and lack of encoded virus genes, which reduce their toxicity and immunogenicity. Interestingly, lentivectors appear to be less genotoxic, clearly a very desirable characteristic for their application in human therapy [90, 102]. Nevertheless, the major advantage of lentivectors is their capacity for transducing postmitotic cells.

However, there are limitations for their use, many of them shared with their γ-retrovirus counterparts. These include the relatively low titers achieved from producer/packaging cells, although they can reach up to 10^7 particles per ml by transient transfection. Even so, lentivector preparations require concentration and further purification for their application in human therapy.

2.4 Summary and Conclusions

Due to their particular biology and the scientific effort invested in retrovirus research, gene vectors based on retroviruses were amongst the first to be used in gene therapy. They have been very useful in many areas of biomedical research, and in fact, they were the first to be used in successful human gene therapy. However, they present important limitations including serious genotoxic effects. Fortunately, lentivectors derived from HIV-1 seem to be more efficient and less genotoxic. The success of lentivectors in two human gene therapy clinical trials, and lack of genotoxicity in a third indicate that they can be at least as efficient as their γ-retrovirus vector counterparts.

Acknowledgments David Escors is funded by an Arthritis Research UK Career Development Fellowship (18433). Holly Stephenson is funded by the Biomedical Research Centre, Institute of Child Health, UCL. Karine Breckpot is funded by the Fund for Scientific Research-Flandes. The Oxford Structural Genomics Consortium is a registered UK charity (number 1097737) that receives funds from the Canadian Institutes of Health Research, The Canadian Foundation for Innovation, Genome Canada through the Ontario Genomics Institute, GlaxoSmithKline, Karolinska Institutet, the Knut and Alice Wallenberg Foundations, the Ontario Innovation Trust, the Ontario Ministry for Research and Innovation, Merck & Co., Inc., the Novartis Research Foundation, the Swedish Foundation for Strategic Research and the Wellcome Trust.

References

1. Hanafusa H, Hanafusa T, Rubin H (1963) The defectiveness of Rous sarcoma virus. Proc Natl Acad Sci U S A 49:572–580
2. Kamine J, Buchanan JM (1977) Cell-free synthesis of two proteins unique to RNA of transforming virions of Rous sarcoma virus. Proc Natl Acad Sci U S A 74(5):2011–2015
3. Rapp UR, Todaro C (1978) Generation of new mouse sarcoma viruses in cell culture. Science 201(4358):821–824
4. Ellis RW, Defeo D, Shih TY, Gonda MA, Young HA, Tsuchida N, Lowy DR, Scolnick EM (1981) The p21 src genes of Harvey and Kirsten sarcoma viruses originate from divergent members of a family of normal vertebrate genes. Nature 292(5823):506–511
5. Vennstrom B, Bishop JM (1982) Isolation and characterization of chicken DNA homologous to the two putative oncogenes of avian erythroblastosis virus. Cell 28(1):135–143
6. Vennstrom B, Sheiness D, Zabielski J, Bishop JM (1982) Isolation and characterization of c-myc, a cellular homolog of the oncogene (v-myc) of avian myelocytomatosis virus strain 29. J Virol 42(3):773–779
7. Parada LF, Tabin CJ, Shih C, Weinberg RA (1982) Human EJ bladder carcinoma oncogene is homologue of Harvey sarcoma virus ras gene. Nature 297(5866):474–478
8. Dalla-Favera R, Gelmann EP, Martinotti S, Franchini G, Papas TS, Gallo RC, Wong-Staal F (1982) Cloning and characterization of different human sequences related to the onc gene (v-myc) of avian myelocytomatosis virus (MC29). Proc Natl Acad Sci U S A 79(21):6497–6501

9. Goff SP, D'Eustachio P, Ruddle FH, Baltimore D (1982) Chromosomal assignment of the endogenous proto-oncogene C-abl. Science 218(4579):1317–1319
10. Chen IS, Wilhelmsen KC, Temin HM (1983) Structure and expression of c-rel, the cellular homolog to the oncogene of reticuloendotheliosis virus strain T. J Virol 45(1):104–113
11. Klempnauer KH, Ramsay G, Bishop JM, Moscovici MG, Moscovici C, McGrath JP, Levinson AD (1983) The product of the retroviral transforming gene v-myb is a truncated version of the protein encoded by the cellular oncogene c-myb. Cell 33(2):345–355
12. Rapp UR, Goldsborough MD, Mark GE, Bonner TI, Groffen J, Reynolds FH Jr, Stephenson JR (1983) Structure and biological activity of v-raf, a unique oncogene transduced by a retrovirus. Proc Natl Acad Sci U S A 80(14):4218–4222
13. Franchini G, Gurgo C, Guo HG, Gallo RC, Collalti E, Fargnoli KA, Hall LF, Wong-Staal F, Reitz MS Jr (1987) Sequence of simian immunodeficiency virus and its relationship to the human immunodeficiency viruses. Nature 328(6130):539–543
14. Barre-Sinoussi F, Chermann JC, Rey F, Nugeyre MT, Chamaret S, Gruest J, Dauguet C, Axler-Blin C, Vezinet-Brun F, Rouzioux C, Rozenbaum W, Montagnier L (1983) Isolation of a T-lymphotropic retrovirus from a patient at risk for acquired immune deficiency syndrome (AIDS). Science 220(4599):868–871
15. Popovic M, Sarngadharan MG, Read E, Gallo RC (1984) Detection, isolation, and continuous production of cytopathic retroviruses (HTLV-III) from patients with AIDS and pre-AIDS. Science 224(4648):497–500
16. Clavel F, Guetard D, Brun-Vezinet F, Chamaret S, Rey MA, Santos-Ferreira MO, Laurent AG, Dauguet C, Katlama C, Rouzioux C et al (1986) Isolation of a new human retrovirus from West African patients with AIDS. Science 233(4761):343–346
17. Clavel F, Mansinho K, Chamaret S, Guetard D, Favier V, Nina J, Santos-Ferreira MO, Champalimaud JL, Montagnier L (1987) Human immunodeficiency virus type 2 infection associated with AIDS in West Africa. N Engl J Med 316(19):1180–1185
18. Hutter G, Nowak D, Mossner M, Ganepola S, Mussig A, Allers K, Schneider T, Hofmann J, Kucherer C, Blau O, Blau IW, Hofmann WK, Thiel E (2009) Long-term control of HIV by CCR5 Delta32/Delta32 stem-cell transplantation. N Engl J Med 360(7):692–698
19. Cavazzana-Calvo M, Payen E, Negre O, Wang G, Hehir K, Fusil F, Down J, Denaro M, Brady T, Westerman K, Cavallesco R, Gillet-Legrand B, Caccavelli L, Sgarra R, Maouche-Chretien L, Bernaudin F, Girot R, Dorazio R, Mulder GJ, Polack A, Bank A, Soulier J, Larghero J, Kabbara N, Dalle B, Gourmel B, Socie G, Chretien S, Cartier N, Aubourg P, Fischer A, Cornetta K, Galacteros F, Beuzard Y, Gluckman E, Bushman F, Hacein-Bey-Abina S, Leboulch P (2010) Transfusion independence and HMGA2 activation after gene therapy of human beta-thalassaemia; 1476–4687 (Electronic) 0028-0836 (Linking); Sep 16 2010, pp 318–322
20. Cartier N, Hacein-Bey-Abina S, Bartholomae CC, Veres G, Schmidt M, Kutschera I, Vidaud M, Abel U, Dal-Cortivo L, Caccavelli L, Mahlaoui N, Kiermer V, Mittelstaedt D, Bellesme C, Lahlou N, Lefrere F, Blanche S, Audit M, Payen E, Leboulch P, l'Homme B, Bougneres P, Von Kalle C, Fischer A, Cavazzana-Calvo M, Aubourg P (2009) Hematopoietic stem cell gene therapy with a lentiviral vector in X-linked adrenoleukodystrophy. Science 326(5954):818–823
21. Vogt VM, Simon MN (1999) Mass determination of rous sarcoma virus virions by scanning transmission electron microscopy. J Virol 73(8):7050–7055
22. Katz RA, Skalka AM (1994) The retroviral enzymes. Annu Rev Biochem 63:133–173
23. Jacks T, Power MD, Masiarz FR, Luciw PA, Barr PJ, Varmus HE (1988) Characterization of ribosomal frameshifting in HIV-1 gag–pol expression. Nature 331(6153):280–283
24. Herschhorn A, Hizi A (2010) Retroviral reverse transcriptases. Cell Mol Life Sci 67(16):2717–2747
25. Gilboa E, Mitra SW, Goff S, Baltimore D (1979) A detailed model of reverse transcription and tests of crucial aspects. Cell 18(1):93–100
26. Watanabe S, Temin HM (1982) Encapsidation sequences for spleen necrosis virus, an avian retrovirus, are between the 5′ long terminal repeat and the start of the gag gene. Proc Natl Acad Sci U S A 79(19):5986–5990
27. Charneau P, Alizon M, Clavel F (1992) A second origin of DNA plus-strand synthesis is required for optimal human immunodeficiency virus replication. J Virol 66(5):2814–2820

28. Rattray AJ, Champoux JJ (1989) Plus-strand priming by Moloney murine leukemia virus. The sequence features important for cleavage by RNase H. J Mol Biol 208(3):445–456
29. Frankel AD, Young JA (1998) HIV-1: fifteen proteins and an RNA. Annu Rev Biochem 67:1–25
30. Rimsky L, Hauber J, Dukovich M, Malim MH, Langlois A, Cullen BR, Greene WC (1988) Functional replacement of the HIV-1 rev protein by the HTLV-1 rex protein. Nature 335(6192):738–740
31. Younis I, Green PL (2005) The human T-cell leukemia virus Rex protein. Front Biosci 10:431–445
32. Boxus M, Twizere JC, Legros S, Dewulf JF, Kettmann R, Willems L (2008) The HTLV-1 Tax interactome. Retrovirology 5:76
33. Briggs JA, Grunewald K, Glass B, Forster F, Krausslich HG, Fuller SD (2006) The mechanism of HIV-1 core assembly: insights from three-dimensional reconstructions of authentic virions. Structure 14(1):15–20
34. Briggs JA, Johnson MC, Simon MN, Fuller SD, Vogt VM (2006) Cryo-electron microscopy reveals conserved and divergent features of gag packing in immature particles of Rous sarcoma virus and human immunodeficiency virus. J Mol Biol 355(1):157–168
35. Briggs JA, Wilk T, Welker R, Krausslich HG, Fuller SD (2003) Structural organization of authentic, mature HIV-1 virions and cores. EMBO J 22(7):1707–1715
36. Wang H, Kavanaugh MP, North RA, Kabat D (1991) Cell-surface receptor for ecotropic murine retroviruses is a basic amino-acid transporter. Nature 352(6337):729–731
37. Fisher RA, Bertonis JM, Meier W, Johnson VA, Costopoulos DS, Liu T, Tizard R, Walker BD, Hirsch MS, Schooley RT et al (1988) HIV infection is blocked in vitro by recombinant soluble CD4. Nature 331(6151):76–78
38. Samson M, Libert F, Doranz BJ, Rucker J, Liesnard C, Farber CM, Saragosti S, Lapoumeroulie C, Cognaux J, Forceille C, Muyldermans G, Verhofstede C, Burtonboy G, Georges M, Imai T, Rana S, Yi Y, Smyth RJ, Collman RG, Doms RW, Vassart G, Parmentier M (1996) Resistance to HIV-1 infection in caucasian individuals bearing mutant alleles of the CCR-5 chemokine receptor gene. Nature 382(6593):722–725
39. Zaitseva M, Blauvelt A, Lee S, Lapham CK, Klaus-Kovtun V, Mostowski H, Manischewitz J, Golding H (1997) Expression and function of CCR5 and CXCR4 on human Langerhans cells and macrophages: implications for HIV primary infection. Nat Med 3(12):1369–1375
40. Earp LJ, Delos SE, Park HE, White JM (2005) The many mechanisms of viral membrane fusion proteins. Curr Top Microbiol Immunol 285:25–66
41. Hughson FM (1997) Enveloped viruses: a common mode of membrane fusion? Curr Biol 7(9):R565–R569
42. Lewis PF, Emerman M (1994) Passage through mitosis is required for oncoretroviruses but not for the human immunodeficiency virus. J Virol 68(1):510–516
43. Gallay P, Swingler S, Song J, Bushman F, Trono D (1995) HIV nuclear import is governed by the phosphotyrosine-mediated binding of matrix to the core domain of integrase. Cell 83(4):569–576
44. Dewannieux M, Harper F, Richaud A, Letzelter C, Ribet D, Pierron G, Heidmann T (2006) Identification of an infectious progenitor for the multiple-copy HERV-K human endogenous retroelements. Genome Res 16(12):1548–1556
45. Mi S, Lee X, Li X, Veldman GM, Finnerty H, Racie L, LaVallie E, Tang XY, Edouard P, Howes S, Keith JC Jr, McCoy JM (2000) Syncytin is a captive retroviral envelope protein involved in human placental morphogenesis. Nature 403(6771):785–789
46. Ting CN, Rosenberg MP, Snow CM, Samuelson LC, Meisler MH (1992) Endogenous retroviral sequences are required for tissue-specific expression of a human salivary amylase gene. Genes Dev 6(8):1457–1465
47. Tarlinton RE, Meers J, Young PR (2006) Retroviral invasion of the koala genome. Nature 442(7098):79–81
48. Burns JC, Friedmann T, Driever W, Burrascano M, Yee JK (1993) Vesicular stomatitis virus G glycoprotein pseudotyped retroviral vectors: concentration to very high titer and

efficient gene transfer into mammalian and nonmammalian cells. Proc Natl Acad Sci U S A 90(17):8033–8037
49. Yee JK, Miyanohara A, LaPorte P, Bouic K, Burns JC, Friedmann T (1994) A general method for the generation of high-titer, pantropic retroviral vectors: highly efficient infection of primary hepatocytes. Proc Natl Acad Sci U S A 91(20):9564–9568
50. Akkina RK, Walton RM, Chen ML, Li QX, Planelles V, Chen IS (1996) High-efficiency gene transfer into CD34 + cells with a human immunodeficiency virus type 1-based retroviral vector pseudotyped with vesicular stomatitis virus envelope glycoprotein G. J Virol 70(4):2581–2585
51. Sandrin V, Boson B, Salmon P, Gay W, Negre D, Le Grand R, Trono D, Cosset FL (2002) Lentiviral vectors pseudotyped with a modified RD114 envelope glycoprotein show increased stability in sera and augmented transduction of primary lymphocytes and CD34 + cells derived from human and nonhuman primates. Blood 100(3):823–832
52. Frecha C, Costa C, Negre D, Gauthier E, Russell SJ, Cosset FL, Verhoeyen E (2008) Stable transduction of quiescent T cells without induction of cycle progression by a novel lentiviral vector pseudotyped with measles virus glycoproteins. Blood 112(13):4843–4852
53. Palu G, Parolin C, Takeuchi Y, Pizzato M (2000) Progress with retroviral gene vectors. Rev Med Virol 10(3):185–202
54. Escors D, Breckpot K (2010) Lentiviral vectors in Gene Therapy: their current status and future potential. Arch Immunol Ther Exp 58(2):107–119
55. Stocking C, Loliger C, Kawai M, Suciu S, Gough N, Ostertag W (1988) Identification of genes involved in growth autonomy of hematopoietic cells by analysis of factor-independent mutants. Cell 53(6):869–879
56. Mann R, Mulligan RC, Baltimore D (1983) Construction of a retrovirus packaging mutant and its use to produce helper-free defective retrovirus. Cell 33(1):153–159
57. Pear WS, Nolan GP, Scott ML, Baltimore D (1993) Production of high-titer helper-free retroviruses by transient transfection. Proc Natl Acad Sci U S A 90(18):8392–8396
58. Soneoka Y, Cannon PM, Ramsdale EE, Griffiths JC, Romano G, Kingsman SM, Kingsman AJ (1995) A transient three-plasmid expression system for the production of high titer retroviral vectors. Nucleic Acids Res 23(4):628–633
59. Andreadis S, Palsson BO (1997) Coupled effects of polybrene and calf serum on the efficiency of retroviral transduction and the stability of retroviral vectors. Hum Gene Ther 8(3):285–291
60. Le Doux JM, Davis HE, Morgan JR, Yarmush ML (1999) Kinetics of retrovirus production and decay. Biotechnol Bioeng 63(6):654–662
61. Parks RJ, Chen L, Anton M, Sankar U, Rudnicki MA, Graham FL (1996) A helper-dependent adenovirus vector system: removal of helper virus by Cre-mediated excision of the viral packaging signal. Proc Natl Acad Sci U S A 93(24):13565–13570
62. Earl PL, Cooper N, Wyatt LS, Moss B, Carroll MW (2001) Preparation of cell cultures and vaccinia virus stocks. In Frederick M. Ausube et al (eds), Current protocols in molecular biology, Chapter16, Unit16
63. Ryser MF, Roesler J, Gentsch M, Brenner S (2007) Gene therapy for chronic granulomatous disease. Expert Opin Biol Ther 7(12):1799–1809
64. Howe SJ, Mansour MR, Schwarzwaelder K, Bartholomae C, Hubank M, Kempski H, Brugman MH, Pike-Overzet K, Chatters SJ, de Ridder D, Gilmour KC, Adams S, Thornhill SI, Parsley KL, Staal FJ, Gale RE, Linch DC, Bayford J, Brown L, Quaye M, Kinnon C, Ancliff P, Webb DK, Schmidt M, von Kalle C, Gaspar HB, Thrasher AJ (2008) Insertional mutagenesis combined with acquired somatic mutations causes leukemogenesis following gene therapy of SCID-X1 patients. J Clin Invest 118(9):3143–3150
65. Gaspar HB, Parsley KL, Howe S, King D, Gilmour KC, Sinclair J, Brouns G, Schmidt M, Von Kalle C, Barington T, Jakobsen MA, Christensen HO, Al Ghonaium A, White HN, Smith JL, Levinsky RJ, Ali RR, Kinnon C, Thrasher AJ (2004) Gene therapy of X-linked severe combined immunodeficiency by use of a pseudotyped gammaretroviral vector. Lancet 364(9452):2181–2187

66. Cavazzana-Calvo M, Hacein-Bey S, de Saint Basile G, Gross F, Yvon E, Nusbaum P, Selz F, Hue C, Certain S, Casanova JL, Bousso P, Deist FL, Fischer A (2000) Gene therapy of human severe combined immunodeficiency (SCID)-X1 disease. Science 288(5466):669–672
67. Hacein-Bey-Abina S, Von Kalle C, Schmidt M, McCormack MP, Wulffraat N, Leboulch P, Lim A, Osborne CS, Pawliuk R, Morillon E, Sorensen R, Forster A, Fraser P, Cohen JI, de Saint Basile G, Alexander I, Wintergerst U, Frebourg T, Aurias A, Stoppa-Lyonnet D, Romana S, Radford-Weiss I, Gross F, Valensi F, Delabesse E, Macintyre E, Sigaux F, Soulier J, Leiva LE, Wissler M, Prinz C, Rabbitts TH, Le Deist F, Fischer A; Cavazzana-Calvo M (2003) LMO2-associated clonal T cell proliferation in two patients after gene therapy for SCID-X1. Science 302(5644):415–419
68. Modlich U, Navarro S, Zychlinski D, Maetzig T, Knoess S, Brugman MH, Schambach A, Charrier S, Galy A, Thrasher AJ, Bueren J, Baum C (2009) Insertional transformation of hematopoietic cells by Self-inactivating lentiviral and gammaretroviral vectors. Mol Ther 17(11):1919–1928
69. von Schwedler U, Kornbluth RS, Trono D (1994) The nuclear localization signal of the matrix protein of human immunodeficiency virus type 1 allows the establishment of infection in macrophages and quiescent T lymphocytes. Proc Natl Acad Sci U S A 91(15):6992–6996
70. Uchida N, Sutton RE, Friera AM, He D, Reitsma MJ, Chang WC, Veres G, Scollay R, Weissman IL (1998) HIV, but not murine leukemia virus, vectors mediate high efficiency gene transfer into freshly isolated G0/G1 human hematopoietic stem cells. Proc Natl Acad Sci U S A 95(20):11939–11944
71. Naldini L, Blomer U, Gallay P, Ory D, Mulligan R, Gage FH, Verma IM, Trono D (1996) In vivo gene delivery and stable transduction of nondividing cells by a lentiviral vector. Science 272(5259):263–267
72. Naldini L, Blomer U, Gage FH, Trono D, Verma IM (1996) Efficient transfer, integration, and sustained long-term expression of the transgene in adult rat brains injected with a lentiviral vector. Proc Natl Acad Sci U S A 93(21):11382–11388
73. Blomer U, Naldini L, Kafri T, Trono D, Verma IM, Gage FH (1997) Highly efficient and sustained gene transfer in adult neurons with a lentivirus vector. J Virol 71(9):6641–6649
74. Heinzinger NK, Bukinsky MI, Haggerty SA, Ragland AM, Kewalramani V, Lee MA, Gendelman HE, Ratner L, Stevenson M, Emerman M (1994) The Vpr protein of human immunodeficiency virus type 1 influences nuclear localization of viral nucleic acids in nondividing host cells. Proc Natl Acad Sci U S A 91(15):7311–7315
75. VandenDriessche T, Thorrez L, Naldini L, Follenzi A, Moons L, Berneman Z, Collen D, Chuah MK (2002) Lentiviral vectors containing the human immunodeficiency virus type-1 central polypurine tract can efficiently transduce nondividing hepatocytes and antigen-presenting cells in vivo. Blood 100(3):813–822
76. Korin YD, Zack JA (1998) Progression to the G1b phase of the cell cycle is required for completion of human immunodeficiency virus type 1 reverse transcription in T cells. J Virol 72(4):3161–3168
77. Breckpot K, Escors D, Arce F, Lopes L, Karwacz K, Van Lint S, Keyaerts M, Collins M (2010) HIV-1 lentiviral vector immunogenicity is mediated by Toll-like receptor 3 (TLR3) and TLR7. J Virol 84:5627–5636
78. Ikeda Y, Takeuchi Y, Martin F, Cosset FL, Mitrophanous K, Collins M (2003) Continuous high-titer HIV-1 vector production. Nat Biotechnol 21(5):569–572
79. Gruber A, Kan-Mitchell J, Kuhen KL, Mukai T, Wong-Staal F (2000) Dendritic cells transduced by multiply deleted HIV-1 vectors exhibit normal phenotypes and functions and elicit an HIV-specific cytotoxic T-lymphocyte response in vitro. Blood 96(4):1327–1333
80. Zufferey R, Nagy D, Mandel RJ, Naldini L, Trono D (1997) Multiply attenuated lentiviral vector achieves efficient gene delivery in vivo. Nat Biotechnol 15(9):871–875
81. Dull T, Zufferey R, Kelly M, Mandel RJ, Nguyen M, Trono D, Naldini L (1998) A third-generation lentivirus vector with a conditional packaging system. J Virol 72(11):8463–8471

82. Sune C, Hayashi T, Liu Y, Lane WS, Young RA, Garcia-Blanco MA (1997) CA150, a nuclear protein associated with the RNA polymerase II holoenzyme, is involved in Tat-activated human immunodeficiency virus type 1 transcription. Mol Cell Biol 17(10):6029–6039
83. Kao SY, Calman AF, Luciw PA, Peterlin BM (1987) Anti-termination of transcription within the long terminal repeat of HIV-1 by tat gene product. Nature 330(6147):489–493
84. Sirven A, Pflumio F, Zennou V, Titeux M, Vainchenker W, Coulombel L, Dubart-Kupperschmitt A, Charneau P (2000) The human immunodeficiency virus type-1 central DNA flap is a crucial determinant for lentiviral vector nuclear import and gene transduction of human hematopoietic stem cells. Blood 96(13):4103–4110
85. Huang ZM, Yen TS (1995) Role of the hepatitis B virus posttranscriptional regulatory element in export of intronless transcripts. Mol Cell Biol 15(7):3864–3869
86. Zufferey R, Donello JE, Trono D, Hope TJ (1999) Woodchuck hepatitis virus posttranscriptional regulatory element enhances expression of transgenes delivered by retroviral vectors. J Virol 73(4):2886–2892
87. Schambach A, Bohne J, Baum C, Hermann FG, Egerer L, von Laer D, Giroglou T (2006) Woodchuck hepatitis virus post-transcriptional regulatory element deleted from X protein and promoter sequences enhances retroviral vector titer and expression. Gene Ther 13(7):641–645
88. Knight S, Bokhoven M, Collins M, Takeuchi Y (2010) Effect of the internal promoter on insertional gene activation by lentiviral vectors with an intact HIV long terminal repeat. J Virol 84(9):4856–4859
89. Maruggi G, Porcellini S, Facchini G, Perna SK, Cattoglio C, Sartori D, Ambrosi A, Schambach A, Baum C, Bonini C, Bovolenta C, Mavilio F, Recchia A (2009) Transcriptional enhancers induce insertional gene deregulation independently from the vector type and design. Mol Ther 17(5):851–856
90. Bokhoven M, Stephen SL, Knight S, Gevers EF, Robinson IC, Takeuchi Y, Collins MK (2009) Insertional gene activation by lentiviral and gammaretroviral vectors. J Virol 83(1):283–294
91. Deglon N, Tseng JL, Bensadoun JC, Zurn AD, Arsenijevic Y, Pereira de Almeida L, Zufferey R, Trono D, Aebischer P (2000) Self-inactivating lentiviral vectors with enhanced transgene expression as potential gene transfer system in Parkinson's disease. Hum Gene Ther 11(1):179–190
92. Miyoshi H, Blomer U, Takahashi M, Gage FH, Verma IM (1998) Development of a self-inactivating lentivirus vector. J Virol 72(10):8150–8157
93. Yu SF, von Ruden T, Kantoff PW, Garber C, Seiberg M, Ruther U, Anderson WF, Wagner EF, Gilboa E (1986) Self-inactivating retroviral vectors designed for transfer of whole genes into mammalian cells. Proc Natl Acad Sci U S A 83(10):3194–3198
94. Apolonia L, Waddington SN, Fernandes C, Ward NJ, Bouma G, Blundell MP, Thrasher AJ, Collins MK, Philpott NJ (2007) Stable gene transfer to muscle using non-integrating lentiviral vectors. Mol Ther 15(11):1947–1954
95. Karwacz K, Mukherjee S, Apolonia L, Blundell MP, Bouma G, Escors D, Collins MK, Thrasher AJ (2009) Nonintegrating lentivector vaccines stimulate prolonged T-cell and antibody responses and are effective in tumor therapy. J Virol 83(7):3094–3103
96. Yanez-Munoz RJ, Balaggan KS, MacNeil A, Howe SJ, Schmidt M, Smith AJ, Buch P, MacLaren RE, Anderson PN, Barker SE, Duran Y, Bartholomae C, von Kalle C, Heckenlively JR, Kinnon C, Ali RR, Thrasher AJ (2006) Effective gene therapy with nonintegrating lentiviral vectors. Nat Med 12(3):348–353
97. Philippe S, Sarkis C, Barkats M, Mammeri H, Ladroue C, Petit C, Mallet J, Serguera C (2006) Lentiviral vectors with a defective integrase allow efficient and sustained transgene expression in vitro and in vivo. Proc Natl Acad Sci U S A 103(47):17684–17689
98. Takayama K, Torashima T (2009) Transgene expression in the mouse cerebellar Purkinje cells with a minimal level of integration using long terminal repeat-modified lentiviral vectors. J Neurovirology 15(5–6):371–379
99. Arce F, Rowe HM, Chain B, Lopes L, Collins MK (2009) Lentiviral vectors transduce proliferating dendritic cell precursors leading to persistent antigen presentation and immunization. Mol Ther 17(9):1643–1650

100. Hu B, Dai B, Wang P (2010) Vaccines delivered by integration-deficient lentiviral vectors targeting dendritic cells induces strong antigen-specific immunity. Vaccine 28(41):6675–6683
101. Negri DR, Bona R, Michelini Z, Leone P, Macchia I, Klotman ME, Salvatore M, Cara A (2010) Transduction of human antigen-presenting cells with integrase-defective lentiviral vector enables functional expansion of primed antigen-specific CD8(+) T cells. Hum Gene Ther 21(8):1029–1035
102. Biffi A, Bartolomae CC, Cesana D, Cartier N, Aubourg P, Ranzani M, Cesani M, Benedicenti F, Plati T, Rubagotti E, Merella S, Capotondo A, Sgualdino J, Zanetti G, von Kalle C, Schmidt M, Naldini L, Montini E (2011) Lentiviral vector common integration sites in preclinical models and a clinical trial reflect a benign integration bias and not oncogenic selection. Blood 117(20):5332–5339

Chapter 3
Cell and Tissue Gene Targeting with Lentiviral Vectors

David Escors, Grazyna Kochan, Holly Stephenson and Karine Breckpot

Abstract One of the main advantages of using lentivectors is their capacity to transduce a wide range of cell types, independently from the cell cycle stage. However, transgene expression in certain cell types is sometimes not desirable, either because of toxicity, cell transformation, or induction of transgene-specific immune responses. In other cases, specific targeting of only cancerous cells within a tumor is sought after for the delivery of suicide genes. Consequently, great effort has been invested in developing strategies to control transgene delivery/expression in a cell/tissue-specific manner. These strategies can broadly be divided in three; particle pseudotyping (surface targeting), which entails modification of the envelope glycoprotein (ENV); transcriptional targeting, which utilizes cell-specific promoters and/or inducible promoters; and posttranscriptional targeting, recently applied in lentivectors by introducing sequence targets for cell-specific microRNAs. In this chapter we describe each of these strategies providing some illustrative examples.

D. Escors (✉)
University College London, Rayne Building, 5 University Street, London WC1E 6JF, UK
e-mail: d.escors@ucl.ac.uk

G. Kochan
Oxford Structural Genomics Consortium, University of Oxford, Old Road Campus Research Building, Roosevelt Drive, Headington, Oxford OX3 7DQ, UK

H. Stephenson
Institute of Child Health, University College London, Great Ormond Street, London WC1N 3JH, UK

K. Breckpot
Vrije Universiteit Brussels, Brussels, Belgium

D. Escors et al., *Lentiviral Vectors and Gene Therapy*,
SpringerBriefs in Biochemistry and Molecular Biology,
DOI: 10.1007/978-3-0348-0402-8_3, © The Author(s) 2012

3.1 Introduction

Lentivectors can effectively transduce a wide range of cells [1, 2]. This property allows gene correction of potentially any cell type. On the other hand, in some circumstances transgene expression is desirable in only a limited number of specific cell targets. For example, intravenous lentivector administration results in gene transfer to hepatocytes in mouse models. However, this also leads to transgene expression in professional antigen presenting cells such as plasmacytoid dendritic cells (pDCs). These cells then trigger a transgene-specific immune response that will result in elimination of transgene-expressing hepatocytes [3]. This "collateral transduction" limits the therapeutic efficacy of some gene therapy protocols. Therefore, in this case DC transduction has to be avoided at all costs. In other circumstances, transgene expression in immune cells is therapeutic. For example, expression of particular mitogen activated protein kinase (MAPK) constitutive activators in myeloid DCs can either enhance antitumor immune responses or inhibit immune responses by modulating DC functions [4, 5]. On the other hand, some of these MAPK modulators may favor cell transformation if expressed in poorly differentiated cell types [6–8]. Therefore, restricted transgene delivery to immune cells would increase biosafety. Finally, a transgene may be toxic in a particular cell lineage but only at certain differentiation stages. This is exemplified in the correction by gene therapy of globoid cell leukodystrophy, a lysosomal storage disease caused by inactivating mutations in galactocerebrosidase (GALC) [9]. While GALC expression is highly toxic in early hematopoietic progenitors, it is therapeutic in mature cells from the hematopoietic lineage [9]. This is an interesting case in which specific transgene delivery was achieved according to the cellular differentiation stage.

Hence, there are many circumstances in which specific targeting to cell types and tissues has to be achieved. Therefore, the lentivector tropism has been modulated by many experimental approaches, and here we will focus on the best-known examples.

3.2 Modification of Lentivector Tropism by Pseudotyping (Surface Targeting)

Transgene delivery by lentivectors depends on the recognition of the target cell by ENV, which is followed by entry into the cell. Therefore, the lentivector tropism is first determined by specific binding to cell surface receptors. As discussed in Chap. 2, lentivectors can acquire a wide range of different envelope glycoproteins during budding at the plasma membrane from the producer cell. This process is called pseudotyping because the resulting virions (pseudovirions) exhibit the surface antigenicity provided by a heterologous ENV [10, 11].

HIV-1 ENV can be used for "pseudotyping" lentivectors, although it does not lead to high titer preparations. For this reason, one of the most widely used envelopes for lentivector pseudotyping is the vesicular stomatitis virus glycoprotein (VSV-G) [2, 12–14]. VSV-G pseudotyping exhibits many advantages; firstly, it stabilizes the vector particle, leading to high titer vector preparations, and allows vector concentration by ultracentrifugation due to its stability [15]. Secondly, VSV-G is a pantropic envelope, and confers a very broad host cell range [16]. In fact, it is unclear whether VSV-G binds a specific ubiquitous cell receptor, or binds to phospholipids in the plasma membrane [15, 17, 18].

However, in some cases restriction of lentivector tropism results in safer in vivo gene delivery, and can also enhance the therapeutic effects by reducing the lentivector dose. This is of interest since reaching high titer retrovirus vector preparations is a major difficulty. For this reason, several strategies have been applied to achieve specific transductional targeting by surface modification of ENV as explained below.

3.2.1 Pseudotyping with Heterologous Viral Proteins

The availability of a broad range of existing viral ENVs combined with the capacity of retrovirus/lentivirus vectors to accommodate heterologous ENVs makes this strategy simple and straightforward. These lentivectors should exhibit the same cell/tissue tropism of the virus from which the ENVs originated. The list of available glycoproteins for lentivector pseudotyping is evergrowing [19]. Summarizing, viral glycoproteins from several viral families have been successfully used, including *Retroviridae*, *Baculoviridae*, *Filoviridae*, *Flaviviridae*, *Arenaviridae*, *Rhabdoviridae*, *Paramyxoviridae,* and *Coronaviridae* [19, 20] (Table 3.1). In this section we will provide key examples.

Lentivectors can be easily pseudotyped with γ-retroviral ENVs such as mouse leukemia virus amphotropic (MLV-A), gibbon ape leukemia virus (GALV), and feline endogenous retrovirus (RD114) envelopes [21–23]. These envelopes recognize cellular receptors expressed in a wide range of human cell types, such as phosphate cotransporters Pit2 for MLV A [24], Pit1 for GALV, and the neutral aminoacid transporter RDR for RD114 envelope [22, 25–27]. In particular cases, lentivector pseudotyping requires certain modifications in these ENVs. For GALV and RD114 ENVs, substitution of the cytoplasmic domain by that of the MLV enhances their incorporation [23, 28, 29]. The substitution of RD114 cleavage site with the site specific for HIV protease increases its activity [30]. Lentivectors pseudotyped with γ-retroviral envelopes effectively transduce CD34+ hematopoietic precursor cells, a requirement for the treatment of several human genetic pathologies [31]. In fact, correction of severe combined immunodeficiency (SCID) was achieved with GALV [32] and MLV-A [33] pseudotyped retrovirus vectors. GALV ENV was used again for the correction of X-linked chronic granulomatous disease (CGD) and Wiskott–Aldrich Syndrome [34, 35]. In contrast, correction of

Table 3.1 Some selected examples of virus envelope glycoproteins commonly used for lentivector and retrovirus vector pseudotyping

Family	Glycoprotein (species)	References
Retroviridae	Human T lymphotropic virus (HTLV)-1, maedi-visna virus, gammaretroviruses	[21, 50–53]
Togaviridae	Semliki forest virus (SFV), venezuelan equine encephalitis virus (VEEV), ross river virus (RRV), and sindbis virus	[38, 54–58]
Rhabdoviridae	Vesicular stomatitis virurs, rabies virus, and mokola lyssavirus	[59–61]
Filoviridae	Ebola, marburg virus	[62, 63]
Orthomyxoviridae	Influenza hemagglutinin	[64]
Coronaviridae	Severe acute respiratory syndrome (SARS) coronavirus	[20, 65]
Baculoviridae	Baculovirus	[40, 41]
Paramyxoviridae	Measles virus	[46]
Arenaviridae	Lymphotropic choriomeningitis virus (LCMV)	[42, 43, 66]

X-linked adrenoleukodystrophy and β-thalassaemia was achieved with VSV-G pseudotyped lentivectors because of their superior transduction efficiency [36, 37].

Lentivectors can also be effectively pseudotyped with envelope proteins from more distant virus families (Table 3.1). These include alphavirus envelopes (Ross River virus and Semliki Forest virus) which exhibit specific tropism towards mouse and human dendritic cells [38, 39]; baculovirus gp64, an insect virus envelope which confers high particle stability and transduction efficiency. Lentivectors pseudotyped with gp64 effectively transduce hepatocytes in vivo, but not cells from the hematopoietic lineage (or very poorly) including DCs [40, 41]. This property can be exploited to prevent transgene-specific immune responses. Lymphocytic choriomeningitis (L-CMV) virus ENV pseudotypes transduce cells from the central nervous system (neurons, neuroblasts, and astrocytes), glioma cells, and also insulin secreting β cells [42, 43].

The list of lentivector pseudotypes and their application is long. However, there is one more case worth explaining in detail due to its relevance for T and B cell human gene therapy. Gene modification of naïve, nonactivated B, and T lymphocytes has always been a scientific challenge. Their efficient transduction requires their activation usually with antiCD3/antiCD28 agonistic antibodies, or by pretreatment with cytokines [44]. This activation alters their phenotype and effector functions before they can be transduced. Even VSV-G lentivector pseudotypes transduce nonactivated T cells inefficiently [44]. Interestingly, efficient transduction of naïve, nonactivated human lymphocytes is achieved with measles virus H and F ENV (H/F) pseudotypes [45, 46]. Measles virus H/F binds to SLAM and CD46 leading to efficient virus entry, nuclear transport, and integration [47]. These lentivectors can also transduce some B cell lymphomas particularly resistant to lentivector transduction [48, 49].

All these examples, especially the last one, demonstrate that it is possible to find an adequate ENV pseudotype for any target cell type.

3.2.2 Pseudotyping with Modified Viral Glycoproteins

The binding function and tropism of ENV pseudotypes can also be altered by modification of ENV residues involved in receptor binding. In some cases, their original tropism can be completely abrogated without affecting their fusion activities. Then, other molecules such as antibodies, cytokines, or receptor ligands can provide an alternative binding method.

An example of altering the natural tropism of ENV to achieve specific DC tropism is the introduction of selected mutations in the Sindbis virus envelope proteins E1/E2. E1/E2 binds to heparan sulfate, present in most cell types, and also to DC-SIGN, a DC-specific molecule. While E2 binds to the cell receptor, E1 mediates membrane fusion. Interestingly, E1 fusion activity is independent of E2 binding to the cell receptor [67]. Specific E2 mutations abolished binding to heparan sulfate but not to DC-SIGN. This modification allowed specific lentivector gene transfer to DCs in vivo [58]. The Sindbis E1/E2 envelope system is also susceptible to other targeting strategies. In some cases, E2 binding capacities have been completely abrogated, while providing alternative binding methods alongside E1/E2 pseudotyping. For example, cell-specific antibodies conjugated to E2 conferred specific tropism towards P-glycoprotein-expressing melanoma cells [68], prostate cancer [69], endothelial cells [70], and CD34+ hematopoietic progenitor cells [71]. Strong antibody conjugation was achieved by introducing the ZZ domain of protein A in E2. Incorporation of antibodies or any other surface molecule alongside modified Sindbis ENVs can effectively target lentivectors to specific cell types [72].

A major setback from the Sindbis-based modification strategies is the dependence on endocytosis for pH-dependent fusion to occur. Physical retargeting of lentivectors does not guarantee their endocytosis. Fortunately, pseudotyping with measles virus F/H envelope glycoproteins circumvents this hurdle. While the H subunit mediates cell binding, the F protein triggers pH-independent fusion [73]. Therefore, F/H lentivector pseudotypes can gain access by direct fusion with the plasma membrane [74]. Similarly to the Sindbis virus E1/E2 system, the measles virus H subunit binding residues can be mutated, and bound to different molecules targeting specific ligands. For example, fusion with either the epidermal growth factor (EGF) or with a CD20-specific single-chain antibody resulted in specific lentivector transduction of EGF receptor expressing cells and CD20+ B lymphocytes, respectively [75]. Of note, the authors of this study remark the high B cell transduction efficiency. However, it is possible that the measles virus H/F envelope system itself is the main determinant for B cell modification [46, 47].

Engineering of retargeted envelope proteins by covalent fusion to natural ligands such as cytokines has proved to be a challenge [76]. These strategies have limited success as the inclusion of a ligand usually inhibits viral entry, with some limited exceptions such as the Sindbis and Measles virus envelope systems [76, 77]. One of these examples is the fusion of influenza heamaglutinin with EGF to target retroviral transduction to EGF receptor-expressing cells [78]. To overcome the inhibition of vector entry, sequence targets for cellular proteases such as metalloproteases (MMP) were introduced to release ENV from the fused ligand/antibody. This strategy has also been applied for the targeting of MMP-expressing tumors using retrovirus and lentivirus vectors [79–83].

3.3 Transcriptional Targeting

Selective targeting of transgene expression to specific cell types can be effectively achieved with cell and tissue-specific promoters. In this situation lentivector transduction is not prevented at cell entry, but rather transgene expression is restricted to specific cell types. The large number of endogenous cellular promoters potentially allows targeted expression to any cell type or tissue. In addition, inducible promoters can also be incorporated in lentivector systems, leading to controlled transgene expression by administration of a given drug. These strategies add an additional control point for the development of cell-specific lentivectors.

3.3.1 Cell and Tissue Specific Promoters

Specific cell type expression can be achieved by incorporating promoters active in these specific cells into the lentivectors. Endogenous cellular promoters are in addition less sensitive to promoter silencing [84, 85]. This is key in human gene therapy; silencing of the γ-retroviral promoter and loss of transgene expression could have contributed to patient death in the CGD clinical trial [86]. Using endogenous cellular promoters results in improved stability and longevity of transgene expression in the target cells. Consequently, a wide range of endogenous promoters has been introduced in retrovirus and lentivirus transfer vectors.

Using this approach and sometimes by combining viral enhancers with endogenous promoters, specific gene expression was achieved in a number of cell types and tissues such as erythroid cells [87–89], endothelial cells [90], retinal cells [91, 92], neurons [93, 94], glial cells [89, 95, 96], and several cell types in the hippocampus [97]. Cells of the liver have also been targeted after intravenous lentivector administration with the use of specific promoters which effectively restricted expression to hepatocytes [84, 98]. In this particular case, the benefits of cell-specific gene expression were clearly shown using the albumin promoter, which resulted in long-term transgene expression in rat liver. In contrast, transgene expression with the cytomegalovirus promoter (CMV) was rapidly silenced [84, 98]. Importantly, hepatocyte-specific promoters prevent transgene expression in professional antigen presenting cells, which could raise transgene-specific immune responses. This is exemplified in the correction of mucopolysaccharidosis type I in a mouse model with lentivector gene therapy. This disease is caused by α-L-iduronidase (IDUA) deficiency, which leads to toxic glycosaminoglycan accumulation in a wide range of cells [99]. Its correction relies on expression of IDUA in the liver by intravenous administration of a therapeutic lentivector. However, IDUA is also expressed in antigen presenting cells, limiting the efficacy, and durability of the correction. To prevent this, IDUA expression was controlled by the albumin promoter, resulting in long-term expression in the liver, and minimal transgene-specific immune responses [99].

Cancer cells have also been specifically targeted using "tumour cell-specific promoters". A lentivector containing a metalloprotease-specific promoter was used to express proapoptotic genes Bax and tBID in MMP2-expressing cancer cell lines [100]. The α-fetoprotein promoter was used to deliver suicide genes to hepatocarcinoma cells [101, 102], and the prostate specific antigen (PSA) promoter for targeting prostate cancer cells. In fact, a lentivector delivering the diphtheria toxin A gene under the control of the PSA promoter has been used to eradicate prostate cancer cells in culture and in a mouse tumor model [103].

In other experimental settings, transgene expression is required in immune cells, particularly DCs. DCs comprise a group of specialized professional antigen presenting cells, which regulate, and control immune responses [4, 104, 105]. DC-specific expression has been achieved to induce antitumor immunity using HLA DRα [106] and Dectin-2 promoters [107]. On the other hand, transcriptional targeting to DCs has been applied to achieve immune suppression. For example, transgene-specific tolerance was achieved by lentivector-mediated CD11c promoter-controlled expression in transgenic mice [108]. Specific DC targeting to achieve immunological tolerance widens the application of gene therapy approaches for the treatment of autoimmune diseases and prevention of graft-versus-host disease.

As mentioned above, possibly one of the most complex tissues/organs to target is the central nervous system, exhibiting a high cellular diversity [109]. In this instance, transcriptional targeting has proved to be a reliable technique. Many cellular promoters are effective for expression in neurons, glial, and hippocampus cells, such as the synapsin and synapsin-1 promoters [89, 93, 97], enolase promoter [94], CD44 promoter, glial fibrillary acidic protein, and vimentin promoters [89, 95]. In some of these cases, high and longlasting transgene expression has been achieved [94, 97], while other promoters have been less efficient [95]. In fact, it is often the case that endogenous promoters are not as strong as those of viral origin. To boost endogenous promoters while retaining their cell specificity, researchers have modified particular cell-specific promoters by combination with other promoters or adding enhancers, and artificial transcriptional activators. This is the case for bidirectional promoters in which a minimal CMV promoter is positioned next to the cell-specific promoter leading to

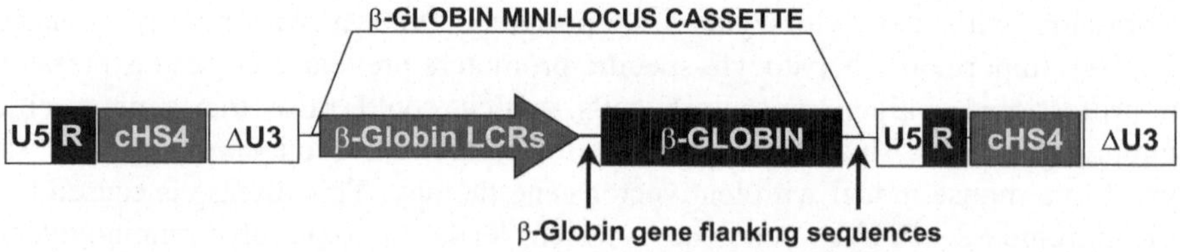

Fig. 3.1 Lentivector design for correction of human β-thalassaemia. The design of the therapeutic lentivector used for the correction of β-thalassaemia is shown. The lentivector contains an expression cassette resembling the endogenous β-globin gene. This includes, apart from the endogenous promoter, the 5′ and 3′ locus control regions (LCRs) placed upstream the β-globin gene [110]. In addition, 5′ and 3′ immediate flanking regions from the endogenous β-globin gene are included and indicated with arrows. Additionally, chicken hypersensitive site 4 (cHS4) β-globin insulator sequences [113] were placed within the LTRs, which prevents silencing of the expression cassette and transcriptional transactivation of adjacent host genes. Please note that this lentivector is a self-inactivating construct (Chap. 2) and the minilocus is placed within the lentivector construct in reverse orientation

transcription in the opposite direction. In this way, transgene expression in the target cells was enhanced [89]. The combination of several promoters within the same construct also allows cell-specific expression of more than one transgene. For example, the interphotoreceptor retinoid binding protein promoter and the guanylate cyclase activating protein promoters were evaluated together with the rhodopsin promoter. These combinations were aimed to achieve specific expression of two trangenes in retinal cells [92].

There is a specific case in which the promoter design has been critical to achieve therapeutic activities in a human gene therapy [37]. Patients suffering from β-thalassaemia contain a nonfunctional allele of β-globin, which results in a marked reduction of its expression. These patients rely on life-long blood transfusions. An obvious approach to correct the disease is to drive β-globin expression in erythroid cells. Although straightforward from a theoretical point of view, the accomplishment of relevant functional β-globin expression has been a challenge. This has been achieved after carefully engineering a lentiviral vector to include the β-globin gene under the transcriptional control of its endogenous promoter, introns, and locus control regions [37, 110–112] (Fig. 3.1).

Summarizing, there is a long list of cell-specific promoters that have been successfully applied in lentiviral vectors, which will surely improve their performance and safety in gene therapy.

3.3.2 Regulatable Promoters

Transgene expression can also be controlled using regulatable promoters. The capacity to regulate transgene expression is crucial for the treatment of genetic diseases for which the timing or levels of expression is critical. A typical example of this is diabetes, in which high blood glucose levels trigger insulin secretion. Many research groups have developed inducible promoters, and many of these systems can be incorporated in lentivectors. Probably, one of the first and most widely used systems utilize tetracycline induction [93, 114–116]. Briefly, there are two main variations of the tetracycline system; tet-on, leading to inducible transgene expression after tetracycline (doxycycline) delivery, and tet-off, which needs constant antibiotic administration to prevent transgene expression [117]. For obvious reasons, the tet-on system is preferred for gene therapy, and most published lentivector systems belong to this category [100, 114, 115, 118, 119], with a few exceptions [93, 120]. In any case, tetracycline-inducible systems are also prone to inactivation and leaky transcription, and their in vivo application is not straightforward [121].

To overcome the disadvantages of tetracycline-dependent inducible systems, other systems have also been adapted to lentivectors, such as the *Drosophila* ecdysone receptor system [122, 123]. This is based on the binding of either ecdysone or synthetic analogs to a heterodimeric protein made of the herpex simplex virus protein VP16 activation domain fused to the ecdysone receptor (VgEcR) and the retinoid X receptor (RXR). VgEcR-RXR then binds to the inducible promoter driving gene transcription [122]. However, this system depends on the administration of multiple lentivector backbones [123]. More recently, it has been successfully reduced to a single lentivector backbone by fusing the tetracycline repressor with the Kruppel-associated Box (KRAB) domain repressor. This novel fusion protein acts as the regulator. This system has achieved tightly regulated conditional transgene expression in the brain, for a drug-inducible transgenic mouse model, or gene silencing in hematopoietic cells [124, 125].

There are quite a number of other inducible systems also adapted to the lentivector system such as the glucocorticoid inducible promoters and mifepristone-inducible systems [126, 127].

3.3.3 Promoters Controlled by Activation State

There are also many promoters upregulated depending on the activation state of different cell types. In most cases, these promoters have been utilized as reporter constructs [4, 128]. An example of these, an NF-κB transactivatable promoter was engineered by fusing NF-κB binding sites upstream of the minimal CMV promoter, driving expression of reporter fluorescent proteins. Addition of toll-like receptor (TLR) agonists such as LPS to DCs modified with these lentivectors resulted in strong transcriptional upregulation of the fluorescent proteins. The interferon β promoter also achieved similar results. These promoters could be useful to express transgenes following DC activation, although they have not been applied in a therapeutic setting yet [4, 128].

3.3.4 Posttranscriptional Targeting

Without any doubt, the discovery of a regulatory system of gene expression based on small noncoding RNAs (microRNAs or miRNAs) has revolutionized biomedical research. These small noncoding 20–24 nt RNAs, termed siRNAs, are partially complementary to a wide range of mRNAs. They can post-transcriptionally inhibit gene expression by either leading to mRNA degradation, translational repression, or mRNA destabilization.

The miRNAs and their activities are regulated by complex mechanisms with many variations depending on the species. Therefore, briefly and oversimplifying, we will describe the main steps controlling miRNA regulation in animal cells. Firstly, active siRNAs are encoded within large precursor RNA molecules called miRNAs (Fig. 3.2) which are transcribed by the RNA polymerase II. These long precursor miRNAs are recognized by a specialized enzymatic pathway (Pasha/ Drosha), which will release the siRNAs in the form of short hairpins (shRNA). The siRNA refers to the hairpin stem together its complementary strand (in some particular cases, the complementary strand can also play regulatory roles). This shRNA is actively exported out of the nucleus to the cytoplasm where it will be recognized by the enzyme complex Dicer (DCR), which will degrade most of the shRNA leaving the stem containing the siRNA target and its complementary sequence (miRNA–miRNA* duplex). This duplex is loaded in the AGO complex (Argonaut), forming the preRISC (RNA Interference Silencing Complex). Subsequently, the miRNA strand is degraded, leaving its complementary miRNA* intact within the RISC complex. The RISC complex will scan mRNAs and when "sufficient" complementation is found between the target mRNA and the miRNA* strand, the mRNA will be degraded. In some cases, the poly-A tail is removed, leading to mRNA destabilization. Alternatively, mRNA translation may be stalled (Fig. 3.2).

So, how has this mechanism been exploited for cell specific targeting of transgene expression? In fact, it is strikingly simple. Different cell types express different patterns of miRNAs, because they are intimately involved in regulation of cell differentiation. Therefore, if a transgene delivered by the lentivector contains a target that is complementary to an endogenously expressed miRNA in cell type A but not cell type B, transgene expression will take place only in cell type B. The transgene mRNA will be degraded in type A alone (Fig. 3.3). This system is called miRNA tagging. However, this is a saturable system. High mRNA levels can saturate RISC complexes, and the mRNA excess will be translated (although resulting in reduced expression levels).

The miRNA tagging technology was quickly applied to solve a major problem in lentivector gene therapy. Direct in vivo lentivector administration leads to rather efficient transgene-specific immune responses, and while this is a desirable characteristic to boost immunity [5, 107, 128–131], it is detrimental for gene therapy of genetic/metabolic disorders. Transgene-specific immune responses dramatically limit the therapeutic activity and survival of corrected cells [3, 99, 132]. To solve

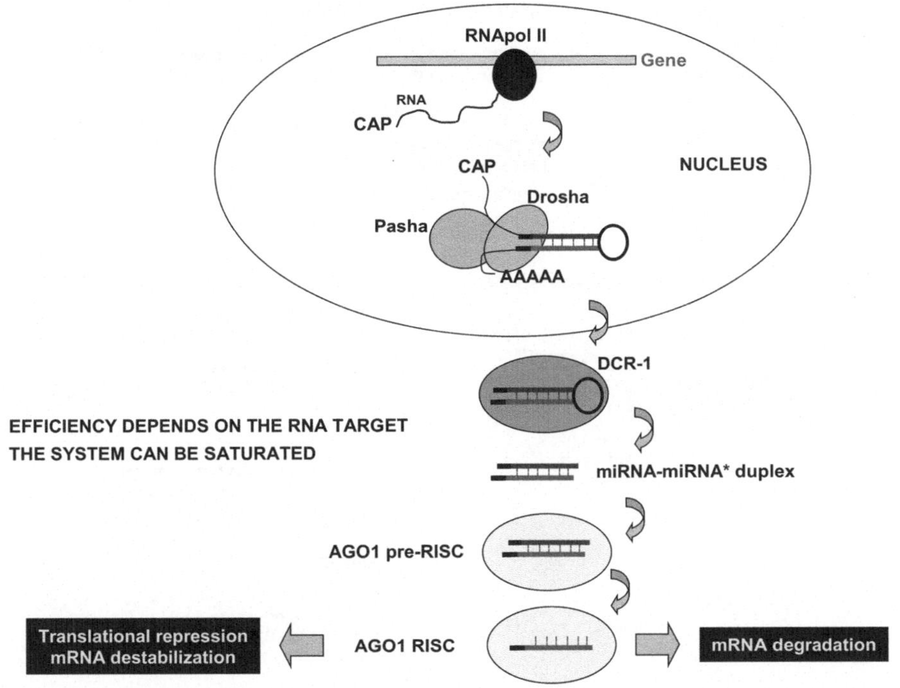

Fig. 3.2 Simplified mechanism of microRNA (mIR) pathways. A simple scheme for the production and function of miRNA-dependent control of gene expression is shown. In the cell nucleus (*upper part*), miRNAs are encoded in large capped RNA molecules transcribed by the cellular RNA polymerase II. These large precursors are recognized by a protein complex containing Drosha and Pasha that will remove the siRNA segment of the short RNA hairpin, which presents a specific stem-loop secondary structure. The shRNA is exported to the cytoplasm and it is bound by Dicer (DCR-1), which will degrade the shRNA leaving a miRNA–miRNA* duplex (miRNA* refers to the complementary sequence to the actual target sequence). This duplex is loaded in a protein complex containing Argonaute (AGO1 pre-RNA Interference Silencing Complex or pre-RISC), and the miRNA strand of the duplex is degraded. The AGO1 RISC contains the complementary strand to the target sequence, which is used to "scan" mRNA molecules exhibiting total and partial complementarity. When matching occurs, the mRNA is either degraded, destabilized or its translation is repressed

this problem, transgene expression was abrogated in cells of the hematopoietic lineage by including four copies of a sequence target for the hematopoietic-specific miRNA 142 3p, downstream of the transgene coding sequence [133]. This strategy ensured that the mRNA encoding the transgene would be degraded only in cells from the hematopoietic lineage, such as lymphocytes, granulocytes and more importantly, macrophages, and DCs. Consequently, intravenous administration of 142 3p-tagged lentivectors resulted in lack of transgene-specific immune responses and sustained, long-term transgene expression in hepatocytes [133]. Interestingly, this strategy resulted in transgene-specific tolerance, as shown by expansion of Foxp3+ regulatory CD4 T cells (Tregs) [134]. Curiously, detargeting antigen

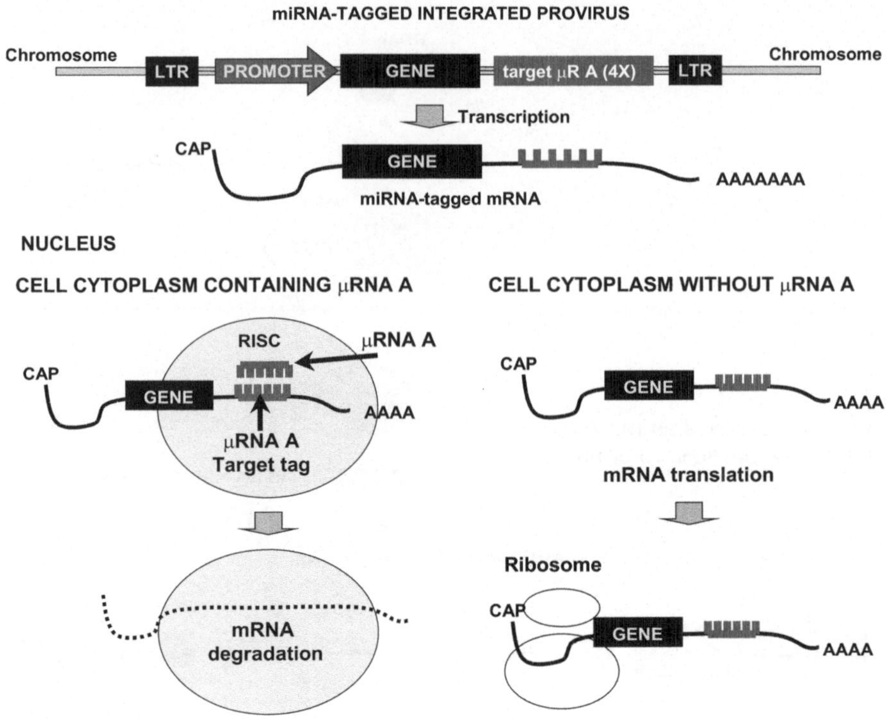

Fig. 3.3 Mechanism of action of miRNA tagging applied to lentivectors. The upper *panel* shows a simplified scheme of a chromosomal integrated lentivector in its "pro-virus" form. This lentivector contains four copies of the target sequence for an ideal miRNA (target μR A, shown as a *red caption*). After transcription, an mRNA is produced encoding the gene of interest followed by the miRNA target sequences. The mRNA is transported out of the nucleus to the cellular cytoplasm (*lower panels*). If the cell is expressing the miRNA A (*left panel*), an RNA silencing complex (RISC) containing the miRNA A (*blue comb*) will bind to its complementary target sequence present in the mRNA (*red comb*). This recognition will lead to disruption of gene translation from that particular mRNA, either by degradation (as shown) or inhibition of translation. If the cell does not express the miRNA A, gene expression will occur as normal by translation (*right panel*)

expression in APCs resulted in Treg expansion [134], when in other experimental settings antigen presentation plays a critical role for differentiation and expansion of antigen-specific Tregs [4, 5, 135, 136]. Interestingly, the same authors demonstrated by using the same miRNA-detargeting strategy that transgene expression in hepatocytes was required for immunological tolerance [134]. Detargeting transgene expression using miRNA 142 3p effectively allowed factor IX expression in liver without raising immune responses, leading to correction of hemophilia B in a mouse model [137].

Another application of miRNA tagging is transgene expression corresponding to specific differentiation or activation stages, by utilizing targets for miRNAs with differentiation stage-dependent variable expression levels [138, 139]. For example, transgene expression can be achieved in only immature DCs, or a combination of different miRNA targets can achieve transgene expression in specific cell types within a given tissue [138]. Another example of targeting expression to cells at different differentiation stages is the introduction of the miRNA 126 target sequence. This particular miRNA is expressed in endothelial, and some epithelial cells, in addition to hematopoietic stem cells. This expression pattern was exploited to target expression of GALC to mature cells from the hematopoietic lineage for the correction of globoid cell leukodystrophy [9]. Curiously, GALC expression in hematopoietic stem cells and early progenitors is highly toxic. In contrast, it is therapeutic in mature cells from the hematopoietic lineage [9]. Therefore, to correct the disease, four copies of the miRNA 126 target sequence were placed downstream GALC gene. Consequently, GALC was only expressed in mature hematopoietic cells, leading to disease correction.

Finally, miRNA tagging can also be exploited to track differentiation pathways, utilizing to the expression pattern of reporter genes containing distinct miRNA target sequences [139].

3.4 Conclusions

The three main groups of lentivector targeting strategies have promising therapeutic applications. Surface targeting ensures the specific entry of the therapeutic vector to targeted cells, while the use of specific promoters can restrict transgene expression if transduction of nontarget cells occurs. Finally, miRNA tagging can add another level of control of transgene expression. In fact, the three strategies have been already applied for the treatment of hemophilia A. A baculovirus gp64-pseudotyped lentivector driving expression of factor VIII from the albumin promoter, in combination with miRNA tagging to avoid transgene expression in APCs, was applied in a mouse model of hemophilia A. Strikingly, in this particular case it was not sufficient to prevent factor VIII-specific immune responses even though liver-specific expression was achieved. Macrophage depletion before lentivector administration had to be performed to achieve therapeutic FVIII levels [40]. The results from this experiment are difficult to explain, as miRNA tagging alone was sufficient to correct hemophilia B without inducing FIX-specific immune responses [137]. This last case demonstrates that even though combining several targeting strategies to avoid transgene-specific immune responses looks promising, specific targeting of viral vectors to cells and tissues is still a challenge.

Acknowledgments David Escors is funded by an Arthritis Research UK Career Development Fellowship (18433). Holly Stephenson is funded by the Biomedical Research Centre, Institute of Child Health, UCL. Karine Breckpot is funded by the Fund for Scientific Research-Flandes. The Oxford Structural Genomics Consortium is a registered UK charity (number 1097737) that receives funds from the Canadian Institutes of Health Research, The Canadian Foundation for Innovation, Genome Canada through the Ontario Genomics Institute, GlaxoSmithKline,

Karolinska Institutet, the Knut and Alice Wallenberg Foundations, the Ontario Innovation Trust, the Ontario Ministry for Research and Innovation, Merck & Co., Inc., the Novartis Research Foundation, the Swedish Foundation for Strategic Research and the Wellcome Trust.

References

1. Uchida N, Sutton RE, Friera AM, He D, Reitsma MJ, Chang WC, Veres G, Scollay R, Weissman IL (1998) HIV, but not murine leukemia virus, vectors mediate high efficiency gene transfer into freshly isolated G0/G1 human hematopoietic stem cells. Proc Natl Acad Sci U S A 95(20):11939–11944
2. Naldini L, Blomer U, Gallay P, Ory D, Mulligan R, Gage FH, Verma IM, Trono D (1996) In vivo gene delivery and stable transduction of nondividing cells by a lentiviral vector. Science 272(5259):263–267
3. Brown BD, Sitia G, Annoni A, Hauben E, Sergi Sergi L, Zingale A, Roncarolo MG, Guidotti LG, Naldini L (2006) In vivo administration of lentiviral vectors triggers a type I interferon response that restricts hepatocyte gene transfer and promotes vector clearance. Blood 1:23–61
4. Arce F, Breckpot K, Stephenson H, Karwacz K, Ehrenstein MR, Collins M, Escors D (2011) Selective ERK activation differentiates mouse and human tolerogenic dendritic cells, expands antigen-specific regulatory T cells, and suppresses experimental inflammatory arthritis. Arthritis Rheum 63:84–95
5. Escors D, Lopes L, Lin R, Hiscott J, Akira S, Davis RJ, Collins MK (2008) Targeting dendritic cell signalling to regulate the response to immunisation. Blood 111(6):3050–3061
6. Gire V, Marshall CJ, Wynford-Thomas D (1999) Activation of mitogen-activated protein kinase is necessary but not sufficient for proliferation of human thyroid epithelial cells induced by mutant Ras. Oncogene 18(34):4819–4832
7. Strobeck MW, Okuda M, Yamaguchi H, Schwartz A, Fukasawa K (1999) Morphological transformation induced by activation of the mitogen-activated protein kinase pathway requires suppression of the T-type Ca2+ channel. J Biol Chem 274(22):15694–15700
8. Cowley S, Paterson H, Kemp P, Marshall CJ (1994) Activation of MAP kinase kinase is necessary and sufficient for PC12 differentiation and for transformation of NIH 3T3 cells. Cell 77(6):841–852
9. Gentner B, Visigalli I, Hiramatsu H, Lechman E, Ungari S, Giustacchini A, Schira G, Amendola M, Quattrini A, Martino S, Orlacchio A, Dick JE, Biffi A, Naldini L (2010) Identification of hematopoietic stem cell-specific miRNAs enables gene therapy of globoid cell leukodystrophy. Sci Transl Med 2(58):58–84
10. Eckner RJ, Steeves RA (1972) A classification of the murine leukemia viruses. Neutralization of pseudotypes of Friend spleen focus-forming virus by type-specific murine antisera. J Exp Med 136(4):832–850
11. Cronin J, Zhang XY, Reiser J (2005) Altering the tropism of lentiviral vectors through pseudotyping. Curr Gene Ther 5(4):387–398
12. Yee JK, Friedmann T, Burns JC (1994) Generation of high-titer pseudotyped retroviral vectors with very broad host range. Methods Cell Biol 43(Pt A):99–112
13. Klages N, Zufferey R, Trono D (2000) A stable system for the high-titer production of multiply attenuated lentiviral vectors. Mol Ther 2(2):170–176
14. Naldini L, Blomer U, Gage FH, Trono D, Verma IM (1996) Efficient transfer, integration, and sustained long-term expression of the transgene in adult rat brains injected with a lentiviral vector. Proc Natl Acad Sci U S A 93(21):11382–11388
15. Burns JC, Friedmann T, Driever W, Burrascano M, Yee JK (1993) Vesicular stomatitis virus G glycoprotein pseudotyped retroviral vectors: concentration to very high titer and

efficient gene transfer into mammalian and nonmammalian cells. Proc Natl Acad Sci U S A 90(17):8033–8037
16. Yee JK, Miyanohara A, LaPorte P, Bouic K, Burns JC, Friedmann T (1994) A general method for the generation of high-titer, pantropic retroviral vectors: highly efficient infection of primary hepatocytes. Proc Natl Acad Sci U S A 91(20):9564–9568
17. Coil DA, Miller AD (2004) Phosphatidylserine is not the cell surface receptor for vesicular stomatitis virus. J Virol 78(20):10920–10926
18. Copreni E, Castellani S, Palmieri L, Penzo M, Conese M (2008) Involvement of glycosaminoglycans in vesicular stomatitis virus G glycoprotein pseudotyped lentiviral vector-mediated gene transfer into airway epithelial cells. J Gene Med 10(12):1294–1302
19. Bouard D, Alazard-Dany D, Cosset FL (2009) Viral vectors: from virology to transgene expression. Br J Pharmacol 157(2):153–165
20. Temperton NJ, Chan PK, Simmons G, Zambon MC, Tedder RS, Takeuchi Y, Weiss RA (2005) Longitudinally profiling neutralizing antibody response to SARS coronavirus with pseudotypes. Emerg Infect Dis 11(3):411–416
21. Strang BL, Ikeda Y, Cosset FL, Collins MK, Takeuchi Y (2004) Characterization of HIV-1 vectors with gammaretrovirus envelope glycoproteins produced from stable packaging cells. Gene Ther 11(7):591–598
22. Faix PH, Feldman SA, Overbaugh J, Eiden MV (2002) Host range and receptor binding properties of vectors bearing feline leukemia virus subgroup B envelopes can be modulated by envelope sequences outside of the receptor binding domain. J Virol 76(23):12369–12375
23. Sandrin V, Boson B, Salmon P, Gay W, Negre D, Le Grand R, Trono D, Cosset FL (2002) Lentiviral vectors pseudotyped with a modified RD114 envelope glycoprotein show increased stability in sera and augmented transduction of primary lymphocytes and CD34+ cells derived from human and nonhuman primates. Blood 100(3):823–832
24. Miller DG, Miller AD (1994) A family of retroviruses that utilize related phosphate transporters for cell entry. J Virol 68(12):8270–8276
25. Sommerfelt MA, Weiss RA (1990) Receptor interference groups of 20 retroviruses plating on human cells. Virology 176(1):58–69
26. Kavanaugh MP, Miller DG, Zhang W, Law W, Kozak SL, Kabat D, Miller AD (1994) Cell-surface receptors for gibbon ape leukemia virus and amphotropic murine retrovirus are inducible sodium-dependent phosphate symporters. Proc Natl Acad Sci U S A 91(15): 7071–7075
27. Rasko JE, Battini JL, Gottschalk RJ, Mazo I, Miller AD (1999) The RD114/simian type D retrovirus receptor is a neutral amino acid transporter. Proc Natl Acad Sci U S A 96(5): 2129–2134
28. Marandin A, Dubart A, Pflumio F, Cosset FL, Cordette V, Chapel-Fernandes S, Coulombel L, Vainchenker W, Louache F (1998) Retrovirus-mediated gene transfer into human CD34+ 38low primitive cells capable of reconstituting long-term cultures in vitro and nonobese diabetic-severe combined immunodeficiency mice in vivo. Hum Gene Ther 9(10):1497–1511
29. Christodoulopoulos I, Cannon PM (2001) Sequences in the cytoplasmic tail of the gibbon ape leukemia virus envelope protein that prevent its incorporation into lentivirus vectors. J Virol 75(9):4129–4138
30. Ikeda Y, Takeuchi Y, Martin F, Cosset FL, Mitrophanous K, Collins M (2003) Continuous high-titer HIV-1 vector production. Nat Biotechnol 21(5):569–572
31. Relander T, Johansson M, Olsson K, Ikeda Y, Takeuchi Y, Collins M, Richter J (2005) Gene transfer to repopulating human CD34+ cells using amphotropic-, GALV-, or RD114-pseudotyped HIV-1-based vectors from stable producer cells. Mol Ther 11(3):452–459
32. Gaspar HB, Parsley KL, Howe S, King D, Gilmour KC, Sinclair J, Brouns G, Schmidt M, Von Kalle C, Barington T, Jakobsen MA, Christensen HO, Al Ghonaium A, White HN, Smith JL, Levinsky RJ, Ali RR, Kinnon C, Thrasher AJ (2004) Gene therapy of X-linked severe combined immunodeficiency by use of a pseudotyped gammaretroviral vector. Lancet 364(9452):2181–2187

33. Cavazzana-Calvo M, Hacein-Bey S, de Saint Basile G, Gross F, Yvon E, Nusbaum P, Selz F, Hue C, Certain S, Casanova JL, Bousso P, Deist FL, Fischer A (2000) Gene therapy of human severe combined immunodeficiency (SCID)-X1 disease. Science 288(5466): 669–672
34. Ott MG, Schmidt M, Schwarzwaelder K, Stein S, Siler U, Koehl U, Glimm H, Kuhlcke K, Schilz A, Kunkel H, Naundorf S, Brinkmann A, Deichmann A, Fischer M, Ball C, Pilz I, Dunbar C, Du Y, Jenkins NA, Copeland NG, Luthi U, Hassan M, Thrasher AJ, Hoelzer D, von Kalle C, Seger R, Grez M (2006) Correction of X-linked chronic granulomatous disease by gene therapy, augmented by insertional activation of MDS1-EVI1, PRDM16 or SETBP1. Nat Med 12(4):401–409
35. Boztug K, Schmidt M, Schwarzer A, Banerjee PP, Diez IA, Dewey RA, Bohm M, Nowrouzi A, Ball CR, Glimm H, Naundorf S, Kuhlcke K, Blasczyk R, Kondratenko I, Marodi L, Orange JS, von Kalle C, Klein C (2010) Stem-cell gene therapy for the Wiskott-Aldrich syndrome. N Engl J Med 363(20):1918–1927
36. Cartier N, Hacein-Bey-Abina S, Bartholomae CC, Veres G, Schmidt M, Kutschera I, Vidaud M, Abel U, Dal-Cortivo L, Caccavelli L, Mahlaoui N, Kiermer V, Mittelstaedt D, Bellesme C, Lahlou N, Lefrere F, Blanche S, Audit M, Payen E, Leboulch P, l'Homme B, Bougneres P, Von Kalle C, Fischer A, Cavazzana-Calvo M, Aubourg P (2009) Hematopoietic stem cell gene therapy with a lentiviral vector in X-linked adrenoleukodystrophy. Science 326(5954):818–823
37. Cavazzana-Calvo M, Payen E, Negre O, Wang G, Hehir K, Fusil F, Down J, Denaro M, Brady T, Westerman K, Cavallesco R, Gillet-Legrand B, Caccavelli L, Sgarra R, Maouche-Chretien L, Bernaudin F, Girot R, Dorazio R, Mulder GJ, Polack A, Bank A, Soulier J, Larghero J, Kabbara N, Dalle B, Gourmel B, Socie G, Chretien S, Cartier N, Aubourg P, Fischer A, Cornetta K, Galacteros F, Beuzard Y, Gluckman E, Bushman F, Hacein-Bey-Abina S, Leboulch P (2010) Transfusion independence and HMGA2 activation after gene therapy of human beta-thalassaemia; 1476-4687 (Electronic) 0028-0836 (Linking); Sep 16 2010, pp 318–322
38. Strang BL, Takeuchi Y, Relander T, Richter J, Bailey R, Sanders DA, Collins MK, Ikeda Y (2005) Human immunodeficiency virus type 1 vectors with alphavirus envelope glycoproteins produced from stable packaging cells. J Virol 79(3):1765–1771
39. Lopes L, Dewannieux M, Takeuchi Y, Collins MK (2011) A lentiviral vector pseudotype suitable for vaccine development. J Gene Med 13(3):181–187
40. Matsui H, Hegadorn C, Ozelo M, Burnett E, Tuttle A, Labelle A, McCray PB Jr, Naldini L, Brown B, Hough C, Lillicrap D (2011) A microRNA-regulated and GP64-pseudotyped lentiviral vector mediates stable expression of FVIII in a murine model of Hemophilia A. Mol Ther 19(4):723–730
41. Schauber CA, Tuerk MJ, Pacheco CD, Escarpe PA, Veres G (2004) Lentiviral vectors pseudotyped with baculovirus gp64 efficiently transduce mouse cells in vivo and show tropism restriction against hematopoietic cell types in vitro. Gene Ther 11(3):266–275
42. Kobinger GP, Deng S, Louboutin JP, Vatamaniuk M, Matschinsky F, Markmann JF, Raper SE, Wilson JM (2004) Transduction of human islets with pseudotyped lentiviral vectors. Hum Gene Ther 15(2):211–219
43. Miletic H, Fischer YH, Neumann H, Hans V, Stenzel W, Giroglou T, Hermann M, Deckert M, Von Laer D (2004) Selective transduction of malignant glioma by lentiviral vectors pseudotyped with lymphocytic choriomeningitis virus glycoproteins. Hum Gene Ther 15(11):1091–1100
44. Perro M, Tsang J, Xue SA, Escors D, Cesco-Gaspere M, Pospori C, Gao L, Hart D, Collins M, Stauss H, Morris EC (2010) Generation of multi-functional antigen-specific human T-cells by lentiviral TCR gene transfer. Gene Ther 17(6):721–732
45. Frecha C, Costa C, Negre D, Gauthier E, Russell SJ, Cosset FL, Verhoeyen E (2008) Stable transduction of quiescent T cells without induction of cycle progression by a novel lentiviral vector pseudotyped with measles virus glycoproteins. Blood 112(13):4843–4852

46. Frecha C, Levy C, Cosset FL, Verhoeyen E (2010) Advances in the field of lentivector-based transduction of T and B lymphocytes for gene therapy. Mol Ther 18(10):1748–1757
47. Frecha C, Levy C, Costa C, Negre D, Amirache F, Buckland R, Russell SJ, Cosset FL, Verhoeyen E (2011) Measles Virus Glycoprotein-Pseudotyped Lentiviral Vector-Mediated Gene Transfer into Quiescent Lymphocytes Requires Binding to both SLAM and CD46 Entry Receptors. J Virol 85(12):5975–5985
48. Frecha C, Costa C, Levy C, Negre D, Russell SJ, Maisner A, Salles G, Peng KW, Cosset FL, Verhoeyen E (2009) Efficient and stable transduction of resting B lymphocytes and primary chronic lymphocyte leukemia cells using measles virus gp displaying lentiviral vectors. Blood 114(15):3173–3180
49. Levy C, Frecha C, Costa C, Rachinel N, Salles G, Cosset FL, Verhoeyen E (2010) Lentiviral vectors and transduction of human cancer B cells. Blood 116(3):498–500 (author reply 500)
50. Coskun AK, Sutton RE (2005) Expression of glucose transporter 1 confers susceptibility to human T-cell leukemia virus envelope-mediated fusion. J Virol 79(7):4150–4158
51. Landau NR, Page KA, Littman DR (1991) Pseudotyping with human T-cell leukemia virus type I broadens the human immunodeficiency virus host range. J Virol 65(1):162–169
52. Lewis BC, Chinnasamy N, Morgan RA, Varmus HE (2001) Development of an avian leukosis-sarcoma virus subgroup A pseudotyped lentiviral vector. J Virol 75(19):9339–9344
53. Zeilfelder U, Bosch V (2001) Properties of wild-type, C-terminally truncated, and chimeric maedi-visna virus glycoprotein and putative pseudotyping of retroviral vector particles. J Virol 75(1):548–555
54. Kahl CA, Marsh J, Fyffe J, Sanders DA, Cornetta K (2004) Human immunodeficiency virus type 1-derived lentivirus vectors pseudotyped with envelope glycoproteins derived from Ross River virus and Semliki Forest virus. J Virol 78(3):1421–1430
55. Kolokoltsov AA, Weaver SC, Davey RA (2005) Efficient functional pseudotyping of oncoretroviral and lentiviral vectors by Venezuelan equine encephalitis virus envelope proteins. J Virol 79(2):756–763
56. Kang Y, Stein CS, Heth JA, Sinn PL, Penisten AK, Staber PD, Ratliff KL, Shen H, Barker CK, Martins I, Sharkey CM, Sanders DA, McCray PB Jr, Davidson BL (2002) In vivo gene transfer using a nonprimate lentiviral vector pseudotyped with Ross River Virus glycoproteins. J Virol 76(18):9378–9388
57. Morizono K, Pariente N, Xie Y, Chen IS (2009) Redirecting lentiviral vectors by insertion of integrin-tageting peptides into envelope proteins. J Gene Med 11(7):549–558
58. Yang L, Yang H, Rideout K, Cho T, Joo KI, Ziegler L, Elliot A, Walls A, Yu D, Baltimore D, Wang P (2008) Engineered lentivector targeting of dendritic cells for in vivo immunization. Nat Biotechnol 26(3):326–334
59. Mazarakis ND, Azzouz M, Rohll JB, Ellard FM, Wilkes FJ, Olsen AL, Carter EE, Barber RD, Baban DF, Kingsman SM, Kingsman AJ, O'Malley K, Mitrophanous KA (2001) Rabies virus glycoprotein pseudotyping of lentiviral vectors enables retrograde axonal transport and access to the nervous system after peripheral delivery. Hum Mol Genet 10(19):2109–2121
60. Desmaris N, Bosch A, Salaun C, Petit C, Prevost MC, Tordo N, Perrin P, Schwartz O, de Rocquigny H, Heard JM (2001) Production and neurotropism of lentivirus vectors pseudotyped with lyssavirus envelope glycoproteins. Mol Ther 4(2):149–156
61. Akkina RK, Walton RM, Chen ML, Li QX, Planelles V, Chen IS (1996) High-efficiency gene transfer into CD34+ cells with a human immunodeficiency virus type 1-based retroviral vector pseudotyped with vesicular stomatitis virus envelope glycoprotein G. J Virol 70(4):2581–2585
62. Kobinger GP, Weiner DJ, Yu QC, Wilson JM (2001) Filovirus-pseudotyped lentiviral vector can efficiently and stably transduce airway epithelia in vivo. Nat Biotechnol 19(3): 225–230
63. Sinn PL, Hickey MA, Staber PD, Dylla DE, Jeffers SA, Davidson BL, Sanders DA, McCray PB Jr (2003) Lentivirus vectors pseudotyped with filoviral envelope glycoproteins

transduce airway epithelia from the apical surface independently of folate receptor alpha. J Virol 77(10):5902–5910
64. Nefkens I, Garcia JM, Ling CS, Lagarde N, Nicholls J, Tang DJ, Peiris M, Buchy P, Altmeyer R (2007) Hemagglutinin pseudotyped lentiviral particles: characterization of a new method for avian H5N1 influenza sero-diagnosis. J Clin Virol 39(1):27–33
65. Hofmann H, Hattermann K, Marzi A, Gramberg T, Geier M, Krumbiegel M, Kuate S, Uberla K, Niedrig M, Pohlmann S (2004) S protein of severe acute respiratory syndrome-associated coronavirus mediates entry into hepatoma cell lines and is targeted by neutralizing antibodies in infected patients. J Virol 78(12):6134–6142
66. Beyer WR, Westphal M, Ostertag W, von Laer D (2002) Oncoretrovirus and lentivirus vectors pseudotyped with lymphocytic choriomeningitis virus glycoprotein: generation, concentration, and broad host range. J Virol 76(3):1488–1495
67. Smit JM, Bittman R, Wilschut J (1999) Low-pH-dependent fusion of Sindbis virus with receptor-free cholesterol- and sphingolipid-containing liposomes. J Virol 73(10):8476–8484
68. Morizono K, Xie Y, Ringpis GE, Johnson M, Nassanian H, Lee B, Wu L, Chen IS (2005) Lentiviral vector retargeting to P-glycoprotein on metastatic melanoma through intravenous injection. Nat Med 11(3):346–352
69. Pariente N, Morizono K, Virk MS, Petrigliano FA, Reiter RE, Lieberman JR, Chen IS (2007) A novel dual-targeted lentiviral vector leads to specific transduction of prostate cancer bone metastases in vivo after systemic administration. Mol Ther 15(11):1973–1981
70. Pariente N, Mao SH, Morizono K, Chen IS (2008) Efficient targeted transduction of primary human endothelial cells with dual-targeted lentiviral vectors. J Gene Med 10(3):242–248
71. Liang M, Pariente N, Morizono K, Chen IS (2009) Targeted transduction of CD34+ hematopoietic progenitor cells in nonpurified human mobilized peripheral blood mononuclear cells. J Gene Med 11(3):185–196
72. Ziegler L, Yang L, Joo K, Yang H, Baltimore D, Wang P (2008) Targeting lentiviral vectors to antigen-specific immunoglobulins. Hum Gene Ther 19(9):861–872
73. Earp LJ, Delos SE, Park HE, White JM (2005) The many mechanisms of viral membrane fusion proteins. Curr Top Microbiol Immunol 285:25–66
74. Funke S, Schneider IC, Glaser S, Muhlebach MD, Moritz T, Cattaneo R, Cichutek K, Buchholz CJ (2009) Pseudotyping lentiviral vectors with the wild-type measles virus glycoproteins improves titer and selectivity. Gene Ther 16(5):700–705
75. Funke S, Maisner A, Muhlebach MD, Koehl U, Grez M, Cattaneo R, Cichutek K, Buchholz CJ (2008) Targeted cell entry of lentiviral vectors. Mol Ther 16(8):1427–1436
76. Waehler R, Russell SJ, Curiel DT (2007) Engineering targeted viral vectors for gene therapy. Nat Rev Genet 8(8):573–587
77. Martin F, Kupsch J, Takeuchi Y, Russell S, Cosset FL, Collins M (1998) Retroviral vector targeting to melanoma cells by single-chain antibody incorporation in envelope. Hum Gene Ther 9(5):737–746
78. Hatziioannou T, Delahaye E, Martin F, Russell SJ, Cosset FL (1999) Retroviral display of functional binding domains fused to the amino terminus of influenza hemagglutinin. Hum Gene Ther 10(9):1533–1544
79. Martin F, Neil S, Kupsch J, Maurice M, Cosset F, Collins M (1999) Retrovirus targeting by tropism restriction to melanoma cells. J Virol 73(8):6923–6929
80. Martin F, Chowdhury S, Neil S, Phillipps N, Collins MK (2002) Envelope-targeted retrovirus vectors transduce melanoma xenografts but not spleen or liver. Mol Ther 5(3): 269–274
81. Szecsi J, Drury R, Josserand V, Grange MP, Boson B, Hartl I, Schneider R, Buchholz CJ, Coll JL, Russell SJ, Cosset FL, Verhoeyen E (2006) Targeted retroviral vectors displaying a cleavage site-engineered hemagglutinin (HA) through HA-protease interactions. Mol Ther 14(5):735–744
82. Duerner LJ, Schwantes A, Schneider IC, Cichutek K, Buchholz CJ (2008) Cell entry targeting restricts biodistribution of replication-competent retroviruses to tumour tissue. Gene Ther 15(22):1500–1510

83. Springfeld C, von Messling V, Frenzke M, Ungerechts G, Buchholz CJ, Cattaneo R (2006) Oncolytic efficacy and enhanced safety of measles virus activated by tumor-secreted matrix metalloproteinases. Cancer Res 66(15):7694–7700
84. Oertel M, Rosencrantz R, Chen YQ, Thota PN, Sandhu JS, Dabeva MD, Pacchia AL, Adelson ME, Dougherty JP, Shafritz DA (2003) Repopulation of rat liver by fetal hepatoblasts and adult hepatocytes transduced ex vivo with lentiviral vectors. Hepatology (Baltimore Md), 37(5):994–1005
85. Liu BH, Wang X, Ma YX, Wang S (2004) CMV enhancer/human PDGF-beta promoter for neuron-specific transgene expression. Gene Ther 11(1):52–60
86. Stein S, Ott MG, Schultze-Strasser S, Jauch A, Burwinkel B, Kinner A, Schmidt M, Kramer A, Schwable J, Glimm H, Koehl U, Preiss C, Ball C, Martin H, Gohring G, Schwarzwaelder K, Hofmann WK, Karakaya K, Tchatchou S, Yang R, Reinecke P, Kuhlcke K, Schlegelberger B, Thrasher AJ, Hoelzer D, Seger R, von Kalle C, Grez M (2010) Genomic instability and myelodysplasia with monosomy 7 consequent to EVI1 activation after gene therapy for chronic granulomatous disease. Nat Med 16(2):198–204
87. Hanawa H, Persons DA, Nienhuis AW (2002) High-level erythroid lineage-directed gene expression using globin gene regulatory elements after lentiviral vector-mediated gene transfer into primitive human and murine hematopoietic cells. Hum Gene Ther 13(17):2007–2016
88. Moreau-Gaudry F, Xia P, Jiang G, Perelman NP, Bauer G, Ellis J, Surinya KH, Mavilio F, Shen CK, Malik P (2001) High-level erythroid-specific gene expression in primary human and murine hematopoietic cells with self-inactivating lentiviral vectors. Blood 98(9): 2664–2672
89. Liu B, Paton JF, Kasparov S (2008) Viral vectors based on bidirectional cell-specific mammalian promoters and transcriptional amplification strategy for use in vitro and in vivo. BMC Biotechnol 8:49
90. De Palma M, Venneri MA, Naldini L (2003) In vivo targeting of tumor endothelial cells by systemic delivery of lentiviral vectors. Hum Gene Ther 14(12):1193–1206
91. Miyoshi H, Takahashi M, Gage FH, Verma IM (1997) Stable and efficient gene transfer into the retina using an HIV-based lentiviral vector. Proc Natl Acad Sci U S A 94(19): 10319–10323
92. Semple-Rowland SL, Eccles KS, Humberstone EJ (2007) Targeted expression of two proteins in neural retina using self-inactivating, insulated lentiviral vectors carrying two internal independent promoters. Mol Vis 13:2001–2011
93. Gascon S, Paez-Gomez JA, Diaz-Guerra M, Scheiffele P, Scholl FG (2008) Dual-promoter lentiviral vectors for constitutive and regulated gene expression in neurons. J Neurosci Methods 168(1):104–112
94. Lai Z, Brady RO (2002) Gene transfer into the central nervous system in vivo using a recombinanat lentivirus vector. J Neurosci Res 67(3):363–371
95. Greenberg KP, Geller SF, Schaffer DV, Flannery JG (2007) Targeted transgene expression in muller glia of normal and diseased retinas using lentiviral vectors. Invest Ophthalmol Vis Sci 48(4):1844–1852
96. Geller SF, Ge PS, Visel M, Greenberg KP, Flannery JG (2007) Functional promoter testing using a modified lentiviral transfer vector. Mol Vis 13:730–739
97. Kuroda H, Kutner RH, Bazan NG, Reiser J (2008) A comparative analysis of constitutive and cell-specific promoters in the adult mouse hippocampus using lentivirus vector-mediated gene transfer. J Gene Med 10(11):1163–1175
98. VandenDriessche T, Thorrez L, Naldini L, Follenzi A, Moons L, Berneman Z, Collen D, Chuah MK (2002) Lentiviral vectors containing the human immunodeficiency virus type-1 central polypurine tract can efficiently transduce nondividing hepatocytes and antigen-presenting cells in vivo. Blood 100(3):813–822
99. Di Domenico C, Di Napoli D, Gonzalez YRE, Lombardo A, Naldini L, Di Natale P (2006) Limited transgene immune response and long-term expression of human alpha-L-

iduronidase in young adult mice with mucopolysaccharidosis type I by liver-directed gene therapy. Hum Gene Ther 17(11):1112–1121
100. Seo E, Kim S, Jho EH (2009) Induction of cancer cell-specific death via MMP2 promoterdependent Bax expression. BMB reports 42(4):217–222
101. Uch R, Gerolami R, Faivre J, Hardwigsen J, Mathieu S, Mannoni P, Bagnis C (2003) Hepatoma cell-specific ganciclovir-mediated toxicity of a lentivirally transduced HSV-TkEGFP fusion protein gene placed under the control of rat alpha-fetoprotein gene regulatory sequences. Cancer Gene Ther 10(9):689–695
102. Gerolami R, Uch R, Brechot C, Mannoni P, Bagnis C (2003) Gene therapy of hepatocarcinoma: a long way from the concept to the therapeutical impact. Cancer Gene Ther 10(9):649–660
103. Zheng JY, Chen D, Chan J, Yu D, Ko E, Pang S (2003) Regression of prostate cancer xenografts by a lentiviral vector specifically expressing diphtheria toxin A. Cancer Gene Ther 10(10):764–770
104. Goold HD, Escors D, Conlan TJ, Chakraverty R, Bennett CL (2011) Conventional DC are required for the activation of helper-dependent CD8 T cell responses after cutaneous vaccination with lentiviral vectors. J Immunol 186(8):4565–4572
105. Breckpot K, Escors D (2009) Dendritic cells for active Anti-cancer Immunotherapy: targeting activation pathways through genetic modification. Endocr Metab Immune Disord Drug Targets 9:328–343
106. Cui Y, Golob J, Kelleher E, Ye Z, Pardoll D, Cheng L (2002) Targeting transgene expression to antigen-presenting cells derived from lentivirus-transduced engrafting human hematopoietic stem/progenitor cells. Blood 99(2):399–408
107. Lopes L, Dewannieux M, Gileadi U, Bailey R, Ikeda Y, Whittaker C, Collin MP, Cerundolo V, Tomihari M, Ariizumi K, Collins MK (2008) Immunization with a lentivector that targets tumor antigen expression to dendritic cells induces potent CD8+ and CD4+ T-cell responses. J Virol 82(1):86–95
108. Zhang J, Zou L, Liu Q, Li J, Zhou J, Wang Y, Li N, Liu T, Wei H, Wu M, Wan Y, Wu Y (2009) Rapid generation of dendritic cell specific transgenic mice by lentiviral vectors. Transgenic Res 18(6):921–931
109. Costantini LC, Bakowska JC, Breakefield XO, Isacson O (2000) Gene therapy in the CNS. Gene Ther 7(2):93–109
110. Grosveld F, van Assendelft GB, Greaves DR, Kollias G (1987) Position-independent, high-level expression of the human beta-globin gene in transgenic mice. Cell 51(6):975–985
111. Grosveld F, Greaves D, Philipsen S, Talbot D, Pruzina S, deBoer E, Hanscombe O, Belhumeur P, Hurst J, Fraser P et al (1990) The dominant control region of the human beta-globin domain. Ann NY Acad Sci 612:152–159
112. Pawliuk R, Westerman KA, Fabry ME, Payen E, Tighe R, Bouhassira EE, Acharya SA, Ellis J, London IM, Eaves CJ, Humphries RK, Beuzard Y, Nagel RL, Leboulch P (2001) Correction of sickle cell disease in transgenic mouse models by gene therapy. Science 294(5550):2368–2371
113. Rivella S, Callegari JA, May C, Tan CW, Sadelain M (2000) The cHS4 insulator increases the probability of retroviral expression at random chromosomal integration sites. J Virol 74(10):4679–4687
114. Reiser J, Lai Z, Zhang XY, Brady RO (2000) Development of multigene and regulated lentivirus vectors. J Virol 74(22):10589–10599
115. Vigna E, Cavalieri S, Ailles L, Geuna M, Loew R, Bujard H, Naldini L (2002) Robust and efficient regulation of transgene expression in vivo by improved tetracycline-dependent lentiviral vectors. Mol Ther 5(3):252–261
116. Efrat S, Fusco-DeMane D, Lemberg H, al Emran O, Wang X (1995) Conditional transformation of a pancreatic beta-cell line derived from transgenic mice expressing a tetracycline-regulated oncogene. Proc Natl Acad Sci U S A 92(8):3576–3580
117. Rose SD, MacDonald RJ (1997) Integration of tetracycline regulation into a cell-specific transcriptional enhancer. J Biol Chem 272(8):4735–4739

118. Farson D, Witt R, McGuinness R, Dull T, Kelly M, Song J, Radeke R, Bukovsky A, Consiglio A, Naldini L (2001) A new-generation stable inducible packaging cell line for lentiviral vectors. Hum Gene Ther 12(8):981–997
119. Georgievska B, Jakobsson J, Persson E, Ericson C, Kirik D, Lundberg C (2004) Regulated delivery of glial cell line-derived neurotrophic factor into rat striatum, using a tetracycline-dependent lentiviral vector. Hum Gene Ther 15(10):934–944
120. Blomer U, Naldini L, Kafri T, Trono D, Verma IM, Gage FH (1997) Highly efficient and sustained gene transfer in adult neurons with a lentivirus vector. J Virol 71(9):6641–6649
121. Johansen J, Rosenblad C, Andsberg K, Moller A, Lundberg C, Bjorlund A, Johansen TE (2002) Evaluation of Tet-on system to avoid transgene down-regulation in ex vivo gene transfer to the CNS. Gene Ther 9(19):1291–1301
122. Saez E, Nelson MC, Eshelman B, Banayo E, Koder A, Cho GJ, Evans RM (2000) Identification of ligands and coligands for the ecdysone-regulated gene switch. Proc Natl Acad Sci U S A 97(26):14512–14517
123. Galimi F, Saez E, Gall J, Hoong N, Cho G, Evans RM, Verma IM (2005) Development of ecdysone-regulated lentiviral vectors. Mol Ther 11(1):142–148
124. Szulc J, Wiznerowicz M, Sauvain MO, Trono D, Aebischer P (2006) A versatile tool for conditional gene expression and knockdown. Nat Methods 3(2):109–116
125. Wiznerowicz M, Szulc J, Trono D (2006) Tuning silence: conditional systems for RNA interference. Nat Methods 3(9):682–688
126. Parker DG, Brereton HM, Klebe S, Coster DJ, Williams KA (2009) A steroid-inducible promoter for the cornea. British J Ophthalmol 93(9):1255–1259
127. Sirin O, Park F (2003) Regulating gene expression using self-inactivating lentiviral vectors containing the mifepristone-inducible system. Gene 323:67–77
128. Breckpot K, Escors D, Arce F, Lopes L, Karwacz K, Van Lint S, Keyaerts M, Collins M (2010) HIV-1 lentiviral vector immunogenicity is mediated by Toll-like receptor 3 (TLR3) and TLR7. J Virol 84:5627–5636
129. Karwacz K, Mukherjee S, Apolonia L, Blundell MP, Bouma G, Escors D, Collins MK, Thrasher AJ (2009) Nonintegrating lentivector vaccines stimulate prolonged T-cell and antibody responses and are effective in tumor therapy. J Virol 83(7):3094–3103
130. Dullaers M, Van Meirvenne S, Heirman C, Straetman L, Bonehill A, Aerts JL, Thielemans K, Breckpot K (2006) Induction of effective therapeutic antitumor immunity by direct in vivo administration of lentiviral vectors. Gene Ther 13(7):630–640
131. Palmowski MJ, Lopes L, Ikeda Y, Salio M, Cerundolo V, Collins MK (2004) Intravenous injection of a lentiviral vector encoding NY-ESO-1 induces an effective CTL response. J Immunol 172(3):1582–1587
132. Rossetti M, Gregori S, Hauben E, Brown BD, Sergi LS, Naldini L, Roncarolo MG (2011) HIV-1-derived lentiviral vectors directly activate plasmacytoid dendritic cells, which in turn induce the maturation of myeloid dendritic cells. Hum Gene Ther 22(2):177–188
133. Brown BD, Venneri MA, Zingale A, Sergi Sergi L, Naldini L (2006) Endogenous microRNA regulation suppresses transgene expression in hematopoietic lineages and enables stable gene transfer. Nat Med 12(5):585–591
134. Annoni A, Brown BD, Cantore A, Sergi LS, Naldini L, Roncarolo MG (2009) In vivo delivery of a microRNA-regulated transgene induces antigen-specific regulatory T cells and promotes immunologic tolerance. Blood 114(25):5152–5161
135. Mahnke K, Qian Y, Knop J, Enk AH (2003) Induction of CD4+/CD25+ regulatory T cells by targeting of antigens to immature dendritic cells. Blood 101(12):4862–4869
136. Kretschmer K, Apostolou I, Hawiger D, Khazaie K, Nussenzweig MC, von Boehmer H (2005) Inducing and expanding regulatory T cell populations by foreign antigen. Nat Immunol 6(12):1219–1227
137. Brown BD, Cantore A, Annoni A, Sergi LS, Lombardo A, Della Valle P, D'Angelo A, Naldini L (2007) A microRNA-regulated lentiviral vector mediates stable correction of hemophilia B mice. Blood 110(13):4144–4152

138. Brown BD, Gentner B, Cantore A, Colleoni S, Amendola M, Zingale A, Baccarini A, Lazzari G, Galli C, Naldini L (2007) Endogenous microRNA can be broadly exploited to regulate transgene expression according to tissue, lineage and differentiation state. Nat Biotechnol 25(12):1457–1467
139. Sachdeva R, Jonsson ME, Nelander J, Kirkeby A, Guibentif C, Gentner B, Naldini L, Bjorklund A, Parmar M, Jakobsson J (2011) Tracking differentiating neural progenitors in pluripotent cultures using microRNA-regulated lentiviral vectors. Proc Natl Acad Sci U S A 107(25):11602–11607

Chapter 4
Immunomodulation by Genetic Modification Using Lentiviral Vectors

Frederick Arce, Karine Breckpot, Grazyna Kochan and David Escors

Abstract Modulation of the immune response is key for the prevention and therapy of different pathologic conditions. In the context of infectious diseases or cancer, the aim is to activate or enhance immune responses against infectious agents or tumor-associated antigens (TAAs). This has been traditionally accomplished through protein or peptide vaccines administered in combination with adjuvants. Despite its many successes, as highlighted by the eradication of smallpox for example, vaccination has not yet proved effective for the treatment of a wide variety of conditions. In other cases, such as in autoimmune and allergic diseases, the objective is instead to achieve immunosuppression, or even induce tolerance towards auto antigens or innocuous xenoantigens. So far, most of the existing immunosuppressive strategies are palliative rather than curative and lack specificity. Therefore, alternative immunization strategies are needed, especially for chronic infections or malignant diseases where immune responses must be reactivated. Lentivector gene therapy can offer an alternative to modulate immune responses.

F. Arce
Paul O'Gorman Building, 72 Huntley Street, London WC1E 6BT, UK

K. Breckpot
Vrije Universiteit Brussels, Brussels, Belgium

G. Kochan
Oxford Structural Genomics Consortium, University of Oxford,
Old Road Campus Research Building, Roosevelt Drive, Headington, Oxford OX3 7DQ, UK

D. Escors (✉)
Rayne Institute, University College London, 5 University Street, London WC1E 6JF, UK
e-mail: d.escors@ucl.ac.uk

D. Escors et al., *Lentiviral Vectors and Gene Therapy*,
SpringerBriefs in Biochemistry and Molecular Biology,
DOI: 10.1007/978-3-0348-0402-8_4,

4.1 Introduction to Genetic Immunotherapy

In the last few decades, genetic modification of cells from the immune system has emerged as a possible way to circumvent the limitations of existing immunotherapies. This approach aims to achieve specificity and effectiveness by targeting key cells that control immune responses, or by manipulating intracellular signaling pathways that cannot be directly modulated by traditional therapies.

Genetic immunotherapy has focused primarily on two cell types that are pivotal controllers of the innate and adaptive immune responses, antigen presenting cells (APCs) and effector T lymphocytes. In the first case, genes encoding antigens of interest are specifically delivered to APCs, which will process and present them to T cells on major histocompatibility complexes (MHC). Alternatively, genes encoding modulators of APC functions can also be delivered together with antigen genes, thus modifying the type of cellular response to the antigen.

T lymphocytes can also be directly modified. T cells express on their surface T cell receptors (TCRs) which recognize antigen associated to MHC from APCs. Vectors expressing specific TCRs or chimeric antigen receptors (CARs) can be delivered to T cells. Thus, genetically engineered T cells can afterwards be adoptively transferred into the host, where they exert their effector functions (passive immunotherapy). In contrast to APC modification, this strategy bypasses the need of mounting an immune response in the host and assures that T cells are specific for the desired antigen.

Different genetic vectors have been used for either approach. Each of them has advantages and disadvantages regarding their biosafety, production, antivector immunity, and quality of the immune response that they raise. Among them, retroviral and lentiviral vectors have been extensively used for genetic modification of APCs and T cells with promising results. This chapter will summarize their current use for immunomodulation, focusing on immune stimulation for the treatment of cancer and infectious diseases, and on the more recent studies in immunosuppression for the treatment of autoimmune diseases.

4.2 Lentivector Gene Therapy for Immunization

4.2.1 Lentivectors for Genetic Modification of Dendritic Cells

T cell-mediated cellular immunity is required for the eradication of malignant cancer cells and the control of infections, such as human immunodeficiency (HIV) virus infection, chronic viral hepatitis, and malaria. The initiation of the cellular immune response depends on the interaction of T cells with professional APCs. Dendritic cells (DCs) are the most immunogenic representative of these cells and therefore, much of the research in immunotherapy has focused on them [1].

The regulation of the cellular immune response by DCs depends on the delivery of different signals to receptors present in lymphocytes [2]. Signal 1 is delivered by the specific recognition of the MHC/peptide complex by cognate T cell receptors (TCRs). Effective T cell activation and expansion requires at least second costimulatory signal/signals, usually called signal 2. In addition, DCs also provide a third signal (signal 3) which depends on cytokines and other immunoregulatory molecules that regulated CD8 T cell cytotoxic activities and CD4 T differentiation [3, 4]. Any immunomodulatory strategy should provide the appropriate combination of three signals to achieve the required T cell response, whether it is immunisation or tolerance.

Lentivectors have been extensively used to genetically modify DCs, for their susceptibility to transduction without affecting their functionality [5–13]. Additionally, lentivectors can stably integrate their genome in DCs leading to sustained transgene expression. Most importantly, the transgene product is processed and presented in the context of the MHC I and II. This has been extensively demonstrated with model antigens such ovalbumin (OVA), tumor-associated antigens such as Melan-A, tyrosinase related protein, NY-ESO and many other antigens from infectious agents, including lymphocytic choriomeningitis virus (LCMV), influenza and human immunodeficiency virus (HIV) [14–18]. Lentivector-transduced DCs expressing these antigens induce in vitro proliferation and activation of both class I or class II-restricted T cell lines or transgenic lymphocytes bearing the cognate T cell receptor.

4.2.2 Lentivector Immunogenicity

Interestingly, lentivectors lead to effective immunotherapy of infectious disease and cancer by providing the additional signals 2 and 3 although the mechanisms are not fully understood [19–21].

Transgenes delivered by lentivectors and expressed in DCs reach the MHC I as endogenous products. A proportion of these expressed proteins are degraded by the proteasome and immunoproteasome. The resulting peptides can translocate to the endoplasmic reticulum (ER) lumen by the transporter associated with antigen processing (TAP) and other TAP-independent mechanisms. TAP-deficient cells transduced with lentivectors do not present transgene products on MHC I [22].

Most importantly for immunotherapy, the proteins expressed by lentivectors can also reach MHC-II through several pathways depending on protein location and trafficking. Secreted proteins can be taken up directly and enter the endocytic pathway. Membrane-bound proteins can be recycled leading to endosomal localization and processing. However, cytoplasmic proteins are poorly processed into MCH II, although they can still enter this pathway by autophagy [23]. To improve MHC class II processing and peptide loading, several molecular approaches have been devised. One strategy consists of fusing endocytic localization sequences or whole proteins with the transgene, such as lysosomal-associated membrane protein 1

(LAMP-1) or part of the MHC II invariant chain (Ii) [24–26]. This method enhances MHC-II presentation and increases the efficacy of the immune response [27].

4.2.3 *Lentivector Delivery of Signals 2 and 3 for Immunomodulation*

To elicit effective immune responses, lentivectors must provide activation signals to DCs. These signals will induce DC maturation by upregulating the expression of costimulatory molecules such as CD80, CD86, CD40, adhesion molecules such as ICAM-I, and MHC molecules. Although lentivectors are generally regarded as poorly immunogenic, their administration in vivo induces DC maturation. This effect is probably mediated by components of the viral particle, and contaminants from the vector preparation.

Some of the lentivector components can stimulate the innate immune system, such as the single-stranded RNA genome and the double-stranded DNA generated after reverse transcription, ligands for TLR7, and TLR9 respectively. Lentivector administration in vivo induces a rapid and transient type I IFN production [28]. Curiously, in several studies, no changes in cDCs were observed on lentivector transduction or HIV-1 infection [29–31], although other studies are in clear conflict [20, 32, 33]. These discrepancies could be explained by differences in experimental systems, the purity of the lentivector preparation and the amounts of vector used for transduction. In fact, contaminants present from the the process of lentivector preparation can affect their immunostimulatory properties. This is the case of vesicular stomatitis virus glycoprotein (VSV-G)-pseudotyped lentivector preparations, which contain VSV-G tubulo-vesicular structures enclosing plasmids that stimulate TLR9 in vitro, leading to type I IFN production by pDCs. These VSV-G vesicles possessed adjuvant capacities [34].

Even residual foetal calf serum (FCS) in concentrated lentivector preparations can also contribute to immunogenicity. In fact, FCS-specific CD4 T cells could be isolated after in vivo lentivector administration which might provide T cell helper functions [35].

4.2.4 *Delivery of DC Molecular Activators with Lentivectors*

Even though lentivector preparations can provide signals 2 and 3 to trigger immune responses, sometimes they are not strong enough to break tolerance to TAAs. Thus, lentivectors can also codeliver genes that modulate DC activation and maturation. This strategy has been successfully applied by lentivector expression of modulators of intracellular signaling cascades, such as those participating in TLR signaling.

Firstly, TLR signaling can be emulated by expression of adaptor molecules associated with this pathway. For example, BMDC transduction with lentivectors encoding Myd88 or TRIF-1 results in IL-6, IL-12 and IFN-α production, with enhanced cytotoxicity [36]. Secondly, NF-κB activation can also be targeted as it is strongly immunostimulatory. Thus, its specific activation has been achieved by lentivector delivery of the Kaposi sarcoma-associated herpes virus FLICE-like inhibitory protein (vFLIP), a viral NF-κB activator. Lentivector coexpression of vFLIP with OVA resulted in DC maturation, enhanced CD4 and CD8 T cell responses, improved tumor-free survival in a lymphoma mouse model and reduction of parasite load after a challenge with OVA-expressing leishmania [37]. NF-κB activation has also been attained by inhibiting negative regulators such as A20, which deactivates adaptor molecules in TLR, TNF, and IL-1 receptor signaling. Lentivector delivery of an A 20-specific short hairpin RNA (shRNA) resulted in upregulation of costimulatory molecules, secretion of proinflammatory cytokines by DCs, improved CD8 T cell responses, and Treg inhibition [38, 39].

In addition, mitogen activated protein kinases (MAPKs) have also been manipulated by the expression of constitutively activated or dominant negative mutants. Constitutive p38 and JNK1 activation resulted in upregulation of some costimulatory molecules, although without significant increase in secretion of proinflammatory cytokines [40–42]. However, coexpression of these MAPK activators with an OVA-containing transgene or Melan-A induced significant antigen-specific CD4 and CD8 T cell responses and improved survival in a murine tumor model for lymphoma. These results have been confirmed in the context of immunization with nonintegrating lentivectors [43].

Expression of other immunostimulatory molecules has also been used to induce DC maturation. CD40 ligand expression activated human DCs by upregulation of CD83, CD80, MHC-I, and IL-12 secretion [44]. This strategy enhanced ex vivo CD4 and CD8 responses against influenza virus epitopes and the tumor-associated antigen gp100. The same authors engineered bone marrow DC precursors ex vivo to generate GM-CSF- and IL-4-expressing DC using lentivectors. Protective and long-lasting immunity against melanoma was achieved when tumor antigens Trp-2 and Mart-1 were expressed by these cells [45].

Some costimulatory signals during antigen presentation can deliver T cell inhibitory signals. Interference with negative costimulation can further enhance T cell responses. Thus, lentivectors encoding shRNAs against programed cell death receptor ligand 1 (PD-L1) in DC resulted in hyperactivated effector T cells by upregulation of Casitas B-lymphoma (Cbl)-b E3 ubiquitin ligase. Interference with PD-L1 costimulation potentiated the immunogenic capacity of DCs and accelerated antitumor immune responses, especially when combined with a p38 activator or an ERK inhibitor [41].

Because of all these reasons, lentivectors are promising immunotherapeutic tools in scenarios where conventional immunization strategies are not effective. This is especially true for the treatment of cancer. There are two major drawbacks for tumor immunotherapy. The first one is that TAAs are usually self-proteins to which there is immunological tolerance. The second one is that tumors themselves are usually strongly immunosuppressive.

There are two possible strategies for immunization using lentiviral vectors. The first one consists of ex vivo DC transduction followed by in vivo administration. Administration of transduced DCs with a lentivector encoding HLA-Cw3 induced proliferation and activation of antigen-specific CD8 T cells in mice [18]. In a similar model, lentivector transduction of DCs rather than peptide pulsing resulted in stronger and longer OVA-specific T cell responses in vivo [46]. Indeed, immunization with lentivector-transduced DCs protected mice from a challenge with OVA-expressing tumor cells and inhibited growth of established tumors. The second strategy, which is cheaper and straightforward, would be direct vaccination with lentivector preparations. This last option has attracted much attention, especially because direct lentivector vaccination achieves DC transduction in the spleen and in lymph nodes after systemic or subcutaneous injections, respectively [12, 16, 47]. Following these studies, the effectiveness of direct administration of lentivectors has been shown by several research groups (Table 1).

4.2.5 Adoptive Cell Transfer of Genetically Modified Lymphocytes

As mentioned before, there are several physiological tolerogenic mechanisms in place that prevent immune responses towards many TAAs, as they are usually identified by the immune system as "self". A major problem is that many TAA-specific T cells have been already eliminated during clonal deletion in the thymus. To circumvent this major drawback, TAA-specific T cells can be engineered and expanded in vitro, followed by adoptive transfer in patients [64]. In fact, clinical benefit has been already achieved in several cancers such as melanoma, synovial cell sarcoma [65], colorectal [65], neuroblastoma [66] or lymphoma [67, 68], although none with lentiviral vectors but rather with retroviral vectors in many cases.

T cell transduction with VSV-G-pseudotyped lentivectors requires some level of T cell stimulation. For example, IL-2 and IL-7 allow efficient lentiviral vector gene transfer and preserve a functional T cell repertoire without skewing the T cell populations [69, 70]. Thus, Wilms tumor antigen (WT1)-specific T cells were engineered using a combination of IL-15 and IL-21 which facilitated WT1 TCR gene transfer. Genetically modified T cells showed redirection of antigen specificity, were multifunctional, produced IL2, IFN-γ and TNF-α while maintaining CD62L and CD28 expression, which are lost in fully differentiated CD8 T cells with poor proliferative potential [71]. Interestingly, in a clinical trial with 15 terminally ill melanoma patients, two of these patients showed direct and complete regression of melanoma. In this case, engineered T cells expressing a MART-1-specific TCR were adoptively transferred [72].

Recently, a major breakthrough was achieved in T cell transduction with lentivectors. Lentivectors pseudotyped with measles virus H/F glycoproteins could efficiently transduce quiescent adult T cells in the absence of any exogenous stimulus, where VSV-G pseudotypes remained refractory. Transduction with

Table 4.1 Effectiveness of direct administration of lentivectors

Antigen	Dose and route	Boosting	Functional analysis	Reference
Cw3 Melan-A	2×10^7 EFU sc	No	Elimination of targets in IVKA[a] Poor secondary response to same LV	[47]
NY-ESO-1	5×10^7 IU iv	NY-ESO-1-VV, day 8	Elimination of targets in IVKA	[16]
Trp2/hsp70 Neu/hsp70	1.6×10^7 PFU sc	No	Decrease in growth rate of small established B16 and G26 tumors Decrease in growth rate of established mammary gland tumors in genetic model	[48]
Full-length HIV-1 Rev/Env Codon-optimized HIV-1 gp120	1×10^7 RT units im	No	In vitro killing of targets	[49]
Melan-A	4×10^6 EFU sc	No	Elimination of targets in IVKA Effective secondary response in challenge with peptide	[50]
OVA	1×10^6–1×10^7 iu sc	OVA-LV, day 150	Elimination of targets in IVKA Decrease in growth rate of established EG.7 tumors	[51]
E-glycoprotein West Nile virus	500 ng p24 ip	No	Protection against challenge with West Nile virus	[52]
OVA	1×10^6 sc	No	Elimination of targets in IVKA Protection against B16-OVA tumor challenge	[53]
OVA	1×10^7 iu iv	OVA –VV, week 3	Complete protection from EG.7 tumor challenge	[27]
HIV-1-derived restricted polyepitopes	0.2–1×10^8 TU ip	No	Elimination of targets in IVKA	[54]

(continued)

Table 4.1 (continued)

Antigen	Dose and route	Boosting	Functional analysis	Reference
Trp2	2 μg p24 iv	Trp2-LV, day 7	Increased survival after B16 tumor challenge, no protection with MHC-II specific promoter	[55]
Codon-optimized HIV-1 gp120 ± GMCSF[b]	1–1.3 × 10^7 RT units im	No	In vitro killing of targets	[56] [57]
NY-ESO-1	4 × 10^6 TU sc	No		[58]
NY-ESO-1	0.001–1 × 10^8 iu sc or 1 × 10^8 iu iv	NY-ESO-1-VV, week 3		[59]
Mutated Trp-1	2.5 × 10^7 TU sc	No	Elimination of targets in IVKA Protection against B16 tumor challenge Elimination of early stage tumors and decrease in growth rate of established tumors	[60]
CEA	0.15 × 10^6 TU sc	CEA-LV, weekly × 3 doses	Regression of CEA-expressing tumors, poor long-term protection	[61]
SIVmac239 gag non-secreted protein[c]	0.25–1 × 10^8 TU sc	Gag-LV with a different envelope, day 79	Protection against intrarrectal SIVmac251 challenge	[62]
OVA[b] Secreted hepatitis B virus surface antigen	150 ng RT sc 1 × 10^7 iu im	No No	Partial regression and increased survival of mice with EG.7 tumors	[43]
hTERT	1 × 10^7 TU sc	Peptide + CFA or DNA		[63]

[a] *IVKA* in vivo killing assay

[b] Immunization with nonintegrating LVs

[c] Study performed in macaques, the rest were done in mice

H/F-pseudotyped lentivectors did not affect cell cycle entry and T cells could maintain the memory and na phenotypes [41, 73–75].

4.3 Lentivector Gene Therapy for the Treatment of Autoimmune Disease

4.3.1 Concept of Tolerance

The organism is constantly in contact with a very wide range of innocuous antigens of different origins. Many are bacterial, others can range from pollen, yeast, mites, and organic/inorganic chemicals. It would not be farfetched to think that the default response of the immune system is tolerance that has to be maintained at all costs. Only when a real threat is apparent, the immune system would strongly react. Consequently, there are several physiological mechanisms in place to establish and maintain tolerance. Possibly one of the most important is the elimination of autoreactive T cells in the thymus by clonal deletion [76]. Nevertheless, clonal deletion itself cannot explain the origin of autoimmune disorders. In these cases, selfantigens are recognized by B and T cells, which mount an immune response with dramatic consequences. Many of these autoreactive T cells that escape from clonal deletion usually differentiate into natural Foxp3+ CD4 regulatory T cells [76, 77]. This T cell lineage is strongly immunosuppressive. There was ample experimental evidence of their existence as early as the 1970s [78–83]. However, their systemic study has only been possible until Sakaguchi and colleagues defined specific markers associated to natural Tregs [84, 85].

Even so, the existence of tolerance towards many innocuous foreign antigens is hard to explain only by clonal deletion and natural Tregs. Usually, the organism gets in touch with these antigens in the periphery, where another Treg cell type differentiates. These Treg cells are termed inducible because they derive from na CD4 T cells. These T cell types can further be classified into Tr1 cells (CD4, CD25, IL10, or TGF-β) and Th3 (CD4, CD25, or Foxp3) [86–89]. For their differentiation, antigens have to be presented in a tolerogenic context, and the DCs performing this tolerogenic antigen presentation are called tolerogenic DCs. Therefore, DCs can also be targeted by gene therapy techniques to differentiate them into DCs with immunosuppressive activities for the treatment of autoimmune disease.

4.3.2 Tolerogenic DCs and Mechanisms of Action

DCs acquire immunosuppressive capacities in specific circumstances after particular stimuli. Antigen presentation by immature DCs leads to T cell anergy, apoptosis, or Treg cell differentiation [90–93]. Additionally, DCs from mucosa and gut are intrinsically tolerogenic as the result of the activity of retinoic acid (vitamin A). Mucosal DCs can also become potently immunosuppressive after contact with some microbial-derived antigens such as TRL2 agonists [94–96]. In fact, ex vivo treatment of DCs with lectin ligands or immunosuppressive cytokines also confers them tolerogenic activities [89, 94, 95, 97–99].

In general, tolerogenic DCs express low levels of MHC and costimulatory molecules such as CD80, CD86, CD83, and ICAM I [21, 42, 99, 100]. However, phenotypically mature DCs can also be potently tolerogenic through secretion of immunosuppressive cytokines [99]. Tolerogenic mechanisms exerted by immunosuppressive DCs are quite different, and it is likely that these are taking place simultaneously. Firstly, antigen presentation by tolerogenic immature DCs can prevent the expansion of effector T cells [101]. Generally speaking, all tolerogenic DCs secrete potent immunosuppressive cytokines during antigen presentation to T cells [21, 94, 95, 98, 100, 102]. For example, if TGF-β is present, na T cells differentiate into antigen-specific Foxp3+ Tregs. Usually, tolerogenic DCs secrete IL-10 which can lead to Tr1 differentiation [102, 103], while keeping DCs in an immature stage [97, 104].

Recent evidence highlights the upregulation of inhibitory costimulatory molecules on the DC surface, which can inhibit T cell activities. One of these is the ligand of the T cell inhibitory receptor PD-1, PD-L1 (or B7-H1) [41, 105]. In fact, PD-L1 upregulation by tumor cells is a key mechanism to avoid immune detection [106]. PD-L1 is expressed ubiquitously, but it is likely that its expression on DCs and other professional antigen presenting cells has a more specific role, especially at the level of regulating T cell activities during antigen presentation [41]. Nevertheless, PD-L1 binding to CD80 on T cells is required for antigen-specific Treg differentiation [107]. A second PD-1 ligand specifically expressed on DCs and macrophages is called PD-L2 or B7-DC, although its role in tolerance is still under debate [108]. Other recently described B7 family members have also been shown to be immunosuppressive [109].

Finally, DCs can be tolerogenic by upregulating the expression of aminoacid-metabolizing enzymes [110]. Increased arginase in APCs suppresses immune responses [111, 112], and indoleamine 2,3-dioxygenase (IDO) upregulation is also potently tolerogenic [113–115].

4.3.3 Genetic Modification of DCs to Induce Immunological Tolerance

The immunosuppressive properties of DCs can also be utilized to induce immunological tolerance for the treatment of autoimmune disease. In this chapter we will put special emphasis on the use of lentivectors for genetic modification of DCs. Lentivectors possess many properties that make them suitable to induce tolerogenic DCs. Thus, DCs can be efficiently reprogramed by expression of immunosuppressive genes with simultaneous delivery of the antigens of interest. In addition, lentivector transduced DCs ensure long-lasting tolerogenic antigen presentation. Finally, as mentioned above, it is not necessary to know specific epitopes for each individual combination of MHC alleles.

The most direct strategy for the generation of tolerogenic DCs is the delivery and expression of potent immunosuppressive cytokines. Retroviral vectors have already been used for the treatment of inflammatory diseases [104, 116, 117]. Similarly, lentivectors expressing IL-10 have been utilized in an OVA-dependent model of experimental asthma [118]. Lentivector-modified tolerogenic DCs could expand IL-10-expressing Foxp3+ Tregs with potent anti-inflammatory properties [118].

As an alternative to direct expression of immunosuppressive cytokines, DCs can be reprogramed by activating tolerogenic sinaling pathways. Thus, specific and sustained MAPK ERK activation induced immune suppression and tolerance when a constitutively active MEK1 mutant was expressed using lentivectors [21, 95, 119–123]. ERK-activated DCs were immature phenotype [21], with CD40 down-modulation and high amount secretion of bioactive TGF-β [21, 100]. ERK-activated DCs could efficiently differentiate antigen-specific Foxp3 Tregs [100], which strongly expanded after a second antigen encounter in inflammatory conditions. This property was used to control antigen-induced inflammatory arthritis in a mouse model [100].

Interestingly, constitutive activation of the type I IFN signaling pathway in DCs using lentivectors was also immunosuppressive. Expression of a constitutively active IRF3 mutant (IRF3 2D), led to high-level expression of IL-10 [21]. IRF3-activated DCs systemically expanded antigen-specific Foxp3 Tregs which resulted in efficient inhibition of effector T cell responses [21]. These results are in agreement with the observation that the type I interferon pathway seems to be immunosuppressive in certain circumstances [124]. For example, IFN-β is used for the treatment of multiple sclerosis [125, 126]. Interestingly, IFN-β and IL-10 secretion follows a common signal transduction pathway [21, 124, 127].

Lentivectors have also been used to inhibit proinflammatory signaling pathways in DCs that in some cases can induce tolerance. Inhibition of the proinflammatory NF-κB pathway is certainly promising [42]. For example, Rel-B silencing by delivery of a specific shRNA [128] was sufficient to prevent DC maturation after TLR triggering. This approach was applied for the treatment of experimental autoimmune myasthenia gravis in mice [128]. Another possibility is the targeted activation of negative feedback mechanisms of proinflammatory pathways. This is the case of overexpression of suppressor of cytokine signaling 3 (SOCS-3) in DCs, which results in immature DCs with impaired proinflammatory signaling [129]. SOCS-3-expressing DCs exhibited reduced expression of IFN-γ, IL-12 and IL-23 with enhanced secretion of IL-10, and could effectively inhibit EAE [129].

Lentiviral vectors can also be applied to suppress autoimmune disorders without the need of targeting the particular pathogenic antigen. This is useful since in many instances these antigens are either unknown or poorly characterized. This was demonstrated for the treatment of experimental collagen-induced arthritis

[130], by administration of a lentivector delivering an siRNA targeted towards B cell activating factor (BAFF) in the inflamed joint [131, 132]. Interestingly, these lentivectors preferentially transduced DCs, interfered with DC maturation, and inhibited Th17 differentiation [130].

Finally, lentivectors can be used for the direct delivery of small peptides with immunosuppressive properties. Direct intraperitoneal vaccination in mice with a lentivector encoding vasointestinal peptide (VIP) inhibited experimental collagen-induced arthritis, inhibited the production of proinflammatory cytokines, and expanded Foxp3+ Tregs [133]. Basically, the same results were obtained by ex vivo transduction of DCs followed by cell transfer. VIP-expressing DCs showed therapeutic activities in a mouse model of EAE and in the cecal ligation and puncture (CLP) model, which are experimental models for multiple sclerosis and sepsis in humans, respectively [134].

4.4 Conclusions

The delivery of immunomodulatory genes together with antigens of interest has made possible the realistic application of gene therapy for the treatment of infectious diseases, cancer, and autoimmune disorders. The efficient and specific targeting of lentivectors to immune cells such as DCs makes them ideal tools for their application in immunomodulation. In addition, the incorporation of antigen-specific T cell receptors into quiescent T cells, opens up the opportunity of designing efficient effector T cells of any given specificity. This is particularly important for the treatment of cancer, in which many of the potential TAA-specific T cells have already been eliminated from the T cell repertoire by clonal deletion.

Acknowledgments David Escors is funded by an Arthritis Research UK Career Development Fellowship (18433). Karine Breckpot is funded by the Fund for Scientific Research-Flandes. The Oxford Structural Genomics Consortium is a registered UK charity (number 1097737) that receives funds from the Canadian Institutes of Health Research, The Canadian Foundation for Innovation, Genome Canada through the Ontario Genomics Institute, GlaxoSmithKline, Karolinska Institutet, the Knut and Alice Wallenberg Foundations, the Ontario Innovation Trust, the Ontario Ministry for Research and Innovation, Merck & Co., Inc., the Novartis Research Foundation, the Swedish Foundation for Strategic Research and the Wellcome Trust.

References

1. Steinman RM, Banchereau J (2007) Nature 449:419–426
2. Janeway CA Jr, Bottomly K (1994) Cell 76:275–285
3. Curtsinger JM, Schmidt CS, Mondino A, Lins DC, Kedl RM, Jenkins MK, Mescher MF (1999) J immunol 162:3256–3262
4. Curtsinger JM, Johnson CM, Mescher MF (2003) J immunol 171:5165–5171
5. Copreni E, Castellani S, Palmieri L, Penzo M, Conese M (2008) J Gene Med 10:1294–1302

6. Yee JK, Friedmann T, Burns JC (1994) Methods Cell Biol 43(Pt A):99–112
7. Esslinger C, Romero P, MacDonald HR (2002) Hum Gene Ther 13:1091–1100
8. Burns JC, Matsubara T, Lozinski G, Yee JK, Friedmann T, Washabaugh CH, Tsonis PA (1994) Dev Biol 165:285–289
9. Bouard D, Alazard-Dany D, Cosset FL (2009) Br J Pharmacol 157:153–165
10. Strang BL, Ikeda Y, Cosset FL, Collins MK, Takeuchi Y (2004) Gene Ther 11:591–598
11. Miller AD, Garcia JV, von Suhr N, Lynch CM, Wilson C, Eiden MV (1991) J Virol 65:2220–2224
12. VandenDriessche T, Thorrez L, Naldini L, Follenzi A, Moons L, Berneman Z, Collen D, Chuah MK (2002) Blood 100:813–822
13. Faix PH, Feldman SA, Overbaugh J, Eiden MV (2002) J Virol 76:12369–12375
14. Miller DG, Miller AD (1994) J Virol 68:8270–8276
15. Lopes L, Fletcher K, Ikeda Y, Collins M (2006) Cancer Immunol Immunother 55:1011–1016
16. Palmowski MJ, Lopes L, Ikeda Y, Salio M, Cerundolo V, Collins MK (2004) J Immunol 172:1582–1587
17. Christodoulopoulos I, Cannon PM (2001) J Virol 75:4129–4138
18. Rasko JE, Battini JL, Gottschalk RJ, Mazo I, Miller AD (1999) Proc Natl Acad Sci U S A 96:2129–2134
19. Goold HD, Escors D, Conlan TJ, Chakraverty R, Bennett CL (2011) J Immunol 186:4565–4572
20. Breckpot K, Escors D, Arce F, Lopes L, Karwacz K, Van Lint S, Keyaerts M, Collins MJ (2010) Virol 84:5627–5636
21. Escors D, Lopes L, Lin R, Hiscott J, Akira S, Davis RJ, Collins MK (2008) Blood 111:3050–3061
22. Zarei S, Abraham S, Arrighi JF, Haller O, Calzascia T, Walker PR, Kundig TM, Hauser C, Piguet V (2004) J Virol 78:7843–7845
23. Paludan C, Schmid D, Landthaler M, Vockerodt M, Kube D, Tuschl T, Munz C (2005) Science 307:593–596
24. Wu TC, Guarnieri FG, Staveley-O'Carroll KF, Viscidi RP, Levitsky HI, Hedrick L, Cho KR, August JT, Pardoll DM (1995) Proc Nat Acad Sci U S A 92:11671–11675
25. Sanderson S, Frauwirth K, Shastri N (1995) Proc Nat Acad Sci U S A 92:7217–7221
26. Gregers TF, Fleckenstein B, Vartdal F, Roepstorff P, Bakke O, Sandlie I (2003) Int Immunol 15:1291–1299
27. Rowe HM, Lopes L, Ikeda Y, Bailey R, Barde I, Zenke M, Chain BM, Collins MK (2006) Mol Ther 13:310–319
28. Brown BD, Sitia G, Annoni A, Hauben E, Sergi LS, Zingale A, Roncarolo MG, Guidotti, LG, Naldini L (2006) Blood
29. Beignon AS, McKenna K, Skoberne M, Manches O, DaSilva I, Kavanagh DG, Larsson M, Gorelick RJ, Lifson JD, Bhardwaj N (2005) J Clin Invest 115:3265–3275
30. Schroers R, Sinha I, Segall H, Schmidt-Wolf IG, Rooney CM, Brenner MK, Sutton RE, Chen SY (2000) Mol Ther 1:171–179
31. Gruber A, Kan-Mitchell J, Kuhen KL, Mukai T, Wong-Staal F (2000) Blood 96:1327–1333
32. Harman AN, Wilkinson J, Bye CR, Bosnjak L, Stern JL, Nicholle M, Lai J, Cunningham AL (2006) J Immunol 177:7103–7113
33. Breckpot K, Emeagi P, Dullaers M, Michiels A, Heirman C, Thielemans K (2007) Hum Gene Ther 18:536–546
34. Pichlmair A, Diebold SS, Gschmeissner S, Takeuchi Y, Ikeda Y, Collins MK, Reise Sousa C (2007) J Virol 81:539–547
35. Bao L, Guo H, Huang X, Tammana S, Wong M, McIvor RS, Zhou X (2009) Gene Ther 16:788–795
36. Akazawa T, Shingai M, Sasai M, Ebihara T, Inoue N, Matsumoto M, Seya T (2007) FEBS Lett 581:3334–3340

37. Rowe HM, Lopes L, Brown N, Efklidou S, Smallie T, Karrar S, Kaye PM, Collins MK (2009) J Virol 83:1555–1562
38. Breckpot K, Aerts-Toegaert C, Heirman C, Peeters U, Beyaert R, Aerts JL, Thielemans K (2009) J Immunol 182:860–870
39. Song XT, Evel-Kabler K, Shen L, Rollins L, Huang XF, Chen SY (2008) Nat Med 14:258–265
40. Escors D, Lopes L, Collins M (2007) Hum Gene Ther 18:191
41. Karwacz K, Bricogne C, Macdonald D, Arce F, Bennett CL, Collins M, Escors D (2011) EMBO Mol Med 3(10):581–592
42. Breckpot K, Escors D (2009) Endocr Metab Immune Disord Drug Targets 9:328–343
43. Karwacz K, Mukherjee S, Apolonia L, Blundell MP, Bouma G, Escors D, Collins MK, Thrasher AJ (2009) J Virol 83:3094–3103
44. Koya RC, Kasahara N, Favaro PM, Lau R, Ta HQ, Weber JS, Stripecke R (2003) J Immunother 26:451–460
45. Koya RC, Kimura T, Ribas A, Rozengurt N, Lawson GW, Faure-Kumar E, Wang HJ, Herschman H, Kasahara N, Stripecke R (2007) Mol Ther 15:971–980
46. He Y, Zhang J, Mi Z, Robbins P, Falo LD Jr (2005) J Immunol 174:3808–3817
47. Esslinger C, Chapatte L, Finke D, Miconnet I, Guillaume P, Levy F, MacDonald HR (2003) J Clin Invest 111:1673–1681
48. Kim JH, Majumder N, Lin H, Watkins S, Falo LD Jr, You Z (2005) Hum Gene Ther 16:1255–1266
49. Buffa V, Negri DR, Leone P, Bona R, Borghi M, Bacigalupo I, Carlei D, Sgadari C, Ensoli B, Cara A (2006) J Gene Virol 87:1625–1634
50. Chapatte L, Colombetti S, Cerottini JC, Levy F (2006) Cancer Res 66:1155–1160
51. Dullaers M, Van Meirvenne S, Heirman C, Straetman L, Bonehill A, Aerts JL, Thielemans K, Breckpot K (2006) Gene Ther 13:630–640
52. Iglesias MC, Frenkiel MP, Mollier K, Souque P, Despres P, Charneau P (2006) J Gene Med 8:265–274
53. He Y, Zhang J, Donahue C, Falo LD Jr (2006) Immunity 24:643–656
54. Iglesias MC, Mollier K, Beignon AS, Souque P, Adotevi O, Lemonnier F, Charneau P (2007) Mol Ther 15:1203–1210
55. Kimura T, Koya RC, Anselmi L, Sternini C, Wang HJ, Comin-Anduix B, Prins RM, Faure-Kumar E, Rozengurt N, Cui Y, Kasahara N, Stripecke R (2007) Mol Ther 15:1390–1399
56. Negri DR, Michelini Z, Baroncelli S, Spada M, Vendetti S, Buffa V, Bona R, Leone P, Klotman ME, Cara A (2007) Mol.Ther 15:1716–1723
57. Negri DR, Michelini Z, Baroncelli S, Spada M, Vendetti S, Bona R, Leone P, Klotman ME, Cara A (2010) J Biomed Biotechnol 2010:534501
58. Casado Garcia, Janda J, Wei J, Chapatte L, Colombetti S, Alves P, Ritter G, Ayyoub M, Valmori D, Chen W, Levy F (2008) Eur J Immunol 38:1867–1876
59. Lopes L, Dewannieux M, Gileadi U, Bailey R, Ikeda Y, Whittaker C, Collin MP, Cerundolo V, Tomihari M, Ariizumi K, Collins MK (2008) J Virol 82:86–95
60. Liu Y, Peng Y, Mi M, Guevara-Patino J, Munn DH, Fu N, He Y (2009) J Immunol 182:5960–5969
61. Loisel-Meyer S, Felizardo T, Mariotti J, Mossoba ME, Foley JE, Kammerer R, Mizue N, Keefe R, McCart JA, Zimmermann W, Dropulic B, Fowler DH, Medin JA (2009) Mol Cancer Ther 8:692–702
62. Beignon AS, Mollier K, Liard C, Coutant F, Munier S, Riviere J, Souque P, Charneau P (2009) J Virol 83:10963–10974
63. Rusakiewicz S, Dosset M, Mollier K, Souque P, Charneau P, Wain-Hobson S, Langlade-Demoyen P, Adotevi O (2010) Vaccine 28:6374–6381
64. Park TS, Rosenberg SA, Morgan RA (2011) Trends Biotechnol
65. Parkhurst MR, Yang JC, Langan RC, Dudley ME, Nathan DA, Feldman SA, Davis JL, Morgan RA, Merino MJ, Sherry RM, Hughes MS, Kammula US, Phan GQ, Lim RM, Wank SA, Restifo NP, Robbins PF, Laurencot CM, Rosenberg SA (2011) Mol Ther 19:620–626

66. Pule MA, Savoldo B, Myers GD, Rossig C, Russell HV, Dotti G, Huls MH, Liu E, Gee AP, Mei Z, Yvon E, Weiss HL, Liu H, Rooney CM, Heslop HE, Brenner MK (2008) Nat Med 14:1264–1270
67. Till BG, Jensen MC, Wang J, Chen EY, Wood BL, Greisman HA, Qian X, James SE, Raubitschek A, Forman SJ, Gopal AK, Pagel JM, Lindgren CG, Greenberg PD, Riddell SR, Press OW (2008) Blood 112:2261–2271
68. Kochenderfer JN, Yu Z, Frasheri D, Restifo NP, Rosenberg SA (2010) Blood 116:3875–3886
69. Cavalieri S, Cazzaniga S, Geuna M, Magnani Z, Bordignon C, Naldini L, Bonini C (2003) Blood 102:497–505
70. Ducrey-Rundquist O, Guyader M, Trono D (2002) J Virol 76:9103–9111
71. Perro M, Tsang J, Xue SA, Escors D, Cesco-Gaspere M, Pospori C, Gao L, Hart D, Collins M, Stauss H, Morris EC (2010) Gene Ther 17:721–732
72. Morgan RA, Dudley ME, Wunderlich JR, Hughes MS, Yang JC, Sherry RM, Royal RE, Topalian SL, Kammula US, Restifo NP, Zheng Z, Nahvi A, de Vries CR, Rogers-Freezer LJ, Mavroukakis SA, Rosenberg SA (2006) Science 314:126–129
73. Frecha C, Costa C, Negre D, Gauthier E, Russell SJ, Cosset FL, Verhoeyen E (2008) Blood 112:4843–4852
74. Frecha C, Costa C, Levy C, Negre D, Russell SJ, Maisner A, Salles G, Peng KW, Cosset FL, Verhoeyen E (2009) Blood 114:3173–3180
75. Frecha C, Levy C, Cosset FL, Verhoeyen E (2010) Mol Ther 18:1748–1757
76. Griesemer AD, Sorenson EC, Hardy MA (2010) Transplantation 90:465–474
77. Sakaguchi S, Yamaguchi T, Nomura T, Ono M (2008) Cell 133:775–787
78. Rich RR, Pierce CW (1973) J Exp Med 137:649–659
79. Ha TY, Waksman BH, Treffers HP (1974) J Exp Med 139:13–23
80. Taussig MJ (1974) Nature 248:236–238
81. Polak L, Turk JL (1974) Nature 249:654–656
82. Kirchner H, Chused TM, Herberman RB, Holden HT, Lavrin DH (1974) J Exp Med 139:1473–1487
83. Basten A, Miller JF, Sprent J, Cheers C (1974) J Exp Med 140:199–217
84. Hori S, Nomura T, Sakaguchi S (2003) Science 299:1057–1061
85. Sakaguchi S (2003) J Clin Invest 112:1310–1312
86. Mahnke K, Qian Y, Knop J, Enk AH (2003) Blood 101:4862–4869
87. O'Garra A, Vieira PL, Vieira P, Goldfeld AE (2004) J Clin Invest 114:1372–1378
88. Peng Y, Laouar Y, Li MO, Green EA, Flavell RA (2004) Proc Natl Acad Sci U S A 101:4572–4577
89. Arce F, Breckpot K, Stephenson H, Karwacz K, Ehrenstein MR, Collins M, Escors D (2011) Arthritis Rheum 63:84–95
90. Kretschmer K, Apostolou I, Hawiger D, Khazaie K, Nussenzweig MC, von Boehmer H (2005) Nat Immunol 6:1219–1227
91. Hawiger D, Inaba K, Dorsett Y, Guo M, Mahnke K, Rivera M, Ravetch JV, Steinman RM, Nussenzweig MC (2001) J Exp Med 194:769–779
92. Bonifaz L, Bonnyay D, Mahnke K, Rivera M, Nussenzweig MC, Steinman RM (2002) J Exp Med 196:1627–1638
93. Dhodapkar MV, Steinman RM, Krasovsky J, Munz C, Bhardwaj N (2001) J Exp Med 193:233–238
94. Ilarregui JM, Croci DO, Bianco GA, Toscano MA, Salatino M, Vermeulen ME, Geffner JR, Rabinovich GA (2009) Nat Immunol 10:981–991
95. Dillon S, Agrawal S, Banerjee K, Letterio J, Denning TL, Oswald-Richter K, Kasprowicz DJ, Kellar K, Pare J, van Dyke T, Ziegler S, Unutmaz D, Pulendran B (2006) J Clin Invest 116:916–928
96. Manicassamy S, Ravindran R, Deng J, Oluoch H, Denning TL, Kasturi SP, Rosenthal KM, Evavold BD, Pulendran B (2009) Nat Med 15:401–409
97. Corinti S, Albanesi C, la Sala A, Pastore S, Girolomoni G (2001) J Immunol 166:4312–4318

98. Ghiringhelli F, Puig PE, Roux S, Parcellier A, Schmitt E, Solary E, Kroemer G, Martin F, Chauffert B, Zitvogel L (2005) J Exp Med 202:919–929
99. Rutella S, Danese S, Leone G (2006) Blood 108:1435–1440
100. Arce F, Breckpot K, Stephenson H, Karwacz K, Ehrenstein MR, Collins M, Escors D (2010) Arthritis Rheum. doi:10.1002/art.30099
101. Rothoeft T, Balkow S, Krummen M, Beissert S, Varga G, Loser K, Oberbanscheidt P, van den Boom F, Grabbe S (2006) Eur J Immunol 36:3105–3117
102. Saraiva M, O'Garra A (2010) Nat Rev Immunol 10:170–181
103. Kuhn R, Lohler J, Rennick D, Rajewsky K, Muller W (1993) Cell 75:263–274
104. Takayama T, Nishioka Y, Lu L, Lotze MT, Tahara H, Thomson AW (1998) Transplantation 66:1567–1574
105. Sakuishi K, Apetoh L, Sullivan JM, Blazar BR, Kuchroo VK, Anderson AC (2010) J Exp Med 207:2187–2194
106. Zhang L, Gajewski TF, Kline J (2009) Blood 114:1545–1552
107. Wang L, Pino-Lagos K, de Vries VC, Guleria I, Sayegh MH, Noelle RJ (2008) Proc Natl Acad Sci U S A 105:9331–9336
108. Radhakrishnan S, Cabrera R, Bruns KM, Van Keulen VP, Hansen MJ, Felts SJ, Pease LR (2009) PLoS ONE 4:e5373
109. Sica GL, Choi IH, Zhu G, Tamada K, Wang SD, Tamura H, Chapoval AI, Flies DB, Bajorath J, Chen L (2003) Immunity 18:849–861
110. Belladonna ML, Orabona C, Grohmann U, Puccetti P (2009) Trends Mol Med 15:41–49
111. Munder M (2009) Br J Pharmacol 158:638–651
112. Norian LA, Rodriguez PC, O'Mara LA, Zabaleta J, Ochoa AC, Cella M, Allen PM (2009) Cancer Res 69:3086–3094
113. Mellor AL, Munn DH (2004) Nat Rev Immunol 4:762–774
114. Fallarino F, Vacca C, Orabona C, Belladonna ML, Bianchi R, Marshall B, Keskin DB, Mellor AL, Fioretti MC, Grohmann U, Puccetti P (2002) Int Immunol 14:65–68
115. Cobbold SP, Adams E, Farquhar CA, Nolan KF, Howie D, Lui KO, Fairchild PJ, Mellor AL, Ron D, Waldmann H (2009) Proc Natl Acad Sci U S A 106:12055–12060
116. Lee WC, Zhong C, Qian S, Wan Y, Gauldie J, Mi Z, Robbins PD, Thomson AW, Lu L (1998) Transplantation 66:1810–1817
117. Morita Y, Yang J, Gupta R, Shimizu K, Shelden EA, Endres J, Mule JJ, McDonagh KT, Fox DA (2001) J Clin Invest 107:1275–1284
118. Henry E, Desmet CJ, Garze V, Fievez L, Bedoret D, Heirman C, Faisca P, Jaspar FJ, Gosset P, Jacquet AP, Desmecht D, Thielemans K, Lekeux P, Moser M, Bureau F (2008) J Immunol 181:7230–7242
119. Agrawal A, Dillon S, Denning TL, Pulendran B (2006) J Immunol 176:5788–5796
120. Caparros E, Munoz P, Sierra-Filardi E, Serrano-Gomez D, Puig-Kroger A, Rodriguez-Fernandez JL, Mellado M, Sancho J, Zubiaur M, Corbi AL (2006) Blood 107:3950–3958
121. Anastasaki C, Estep AL, Marais R, Rauen KA, Patton EE (2009) Hum Mol Genet 18:2543–2554
122. Raingeaud J, Whitmarsh AJ, Barrett T, Derijard B, Davis RJ (1996) Mol Cell Biol 16:1247–1255
123. Pages G, Brunet A, L'Allemain G, Pouyssegur J (1994) EMBO J 13:3003–3010
124. Chang EY, Guo B, Doyle SE, Cheng G (2007) J Immunol 178:6705–6709
125. Billiau A (2006) Antiviral Res 71:108–116
126. Comabella M, Imitola J, Weiner HL, Khoury SJ (2002) J Neuroimmunol 126:205–212
127. Hacker H, Redecke V, Blagoev B, Kratchmarova I, Hsu LC, Wang GG, Kamps MP, Raz E, Wagner H, Hacker G, Mann M, Karin M (2006) Nature 439:204–207
128. Zhang Y, Yang H, Xiao B, Wu M, Zhou W, Li J, Li G, Christadoss P (2009) Mol Immunol 46:657–667
129. Li Y, Chu N, Rostami A, Zhang GX (2006) J Immunol 177:1679–1688
130. Lai Kwan Lam Q, King Hung Ko O, Zheng BJ, Lu L (2008) Proc Natl Acad Sci U S A 105:14993–14998

131. Yang M, Sun L, Wang S, Ko KH, Xu H, Zheng BJ, Cao X, Lu L (2010) J Immunol 184:3321–3325
132. Batten M, Groom J, Cachero TG, Qian F, Schneider P, Tschopp J, Browning JL, Mackay F (2000) J Exp Med 192:1453–1466
133. Delgado M, Toscano MG, Benabdellah K, Cobo M, O'Valle F, Gonzalez-Rey E, Martin F (2008) Arthritis Rheum 58:1026–1037
134. Toscano MG, Delgado M, Kong W, Martin F, Skarica M, Ganea D (2010) Mol Ther 18:1035–1045

Chapter 5
Clinical Grade Lentiviral Vectors

Grazyna Kochan, David Escors, Holly Stephenson and Karine Breckpot

Abstract Thirty years of extensive research culminated in the year 2000 with the publication of the first clearly successful human gene therapy clinical trial. The trial corrected X-linked severe combined immunodeficiency (SCID-X1) in children using a therapeutic γ-retrovirus vector. Soon afterwards, the results of several other trials were published. More recently, lentiviral vectors have been used for the correction of human β-thalassaemia and adrenoleukodystrophy. In this chapter, we discuss the production of clinical grade retro and lentivectors for their application in human therapy.

5.1 Introduction

Retroviral and lentiviral vectors have been extensively used in experimental models to correct immunological and genetic disorders. Their therapeutic effectiveness has been demonstrated numerous times, and naturally, their use in human therapy has followed [1–5]. However, there are several hurdles to overcome in order to translate protocols from experimental models into human therapy. Firstly,

G. Kochan
Oxford Structural Genomics Consortium, University of Oxford,
Old Road Campus Research Building, Roosevelt Drive, Headington, Oxford OX3 7DQ, UK

D. Escors (✉)
University College London, Rayne Building, 5 University Street, London WC1E 6JF, UK
e-mail: d.escors@ucl.ac.uk

H. Stephenson
Institute of Child Health, University College London, Great Ormond Street,
London WC1N 3JH, UK

K. Breckpot
Vrije Universiteit Brussels, Brussels, Belgium

D. Escors et al., *Lentiviral Vectors and Gene Therapy*,
SpringerBriefs in Biochemistry and Molecular Biology,
DOI: 10.1007/978-3-0348-0402-8_5,

vector production has to follow a fully specified, controlled manufacturing procedure following good manufacturing practises (GMPs), set up as guidelines by the appropriate regulatory agencies.

Scaling-up production is a key first step. Most experimental models utilize small animals such as mice, so vector doses have to be scaled up for use in humans. The degree of scaling up will depend on the virus titers achieved during production, and this will be a direct consequence of the nature of the viral vector, the type of producer cells used in the process, and the method of virus production.

The purity of vector preparations is a major issue. Preclinical retroviral/lentiviral vectors contain levels of protein and DNA contaminants unacceptable for use in human therapy. In some cases, antibiotics used to maintain producer cells can also contaminate vector preparations. Thus, several purification/concentration steps have to be implemented to remove as much of the contaminants as possible, to comply with the appropriate medical regulatory agencies.

Classically, therapeutic efficacy is the main end point in animal experimental models, while biosafety is at best a secondary issue. However, after the appearance of several leukemia cases in the SCID-X1 human clinical trials, and some other deaths in nonretrovirus-based gene therapies, biosafety has become a critical issue. Therefore, designing vectors to minimize transforming properties, and removal of protransforming contaminants have become a priority. Additionally, absence of replication-competent viruses arising from the production of therapeutic lentivectors has to be demonstrated.

As in any GMP production method, a significant amount of each vector batch has to be used for quality control assays. These include detection of contaminants and replication-competent virus, DNA mobilization assays and microbiological, and physico-chemical quality control amongst other things. This is particularly a major issue when transient transfection is applied for vector production. The development of stable, GMP-grade producer cells should theoretically yield more homogenous preparations. However, the development of stable producer lentivector cell lines is proving to be a technological challenge.

The final vector yield and cost will determine whether gene therapy can be viable for routine human therapy. As a rule of thumb, a yield of about 10–15% should be expected from purification steps. The yield will directly depend on the vector particle stability, which has to withstand the purification procedures, including filtration, chromatography, high salt concentrations, and ultracentrifugation steps.

5.2 Good Manufacturing Practise Guidelines and Clinical Grade Vector Preparations

GMP guidelines comprise general principles that have to be followed during manufacturing, in this case, of pharmaceutical/biomedical products. GMP guidelines are usually established by the appropriate regulatory medical agencies, and therefore they have to comply with the specific local legislation (which may differ between countries). Some examples of these regulatory agencies are the Food and Drug Administration (FDA) in the USA, the World Health Organization GMP (WHO) and the European Union-GMP. These guidelines ensure that the manufacturing process is controlled, well defined, and that all manufacturing instructions are clearly written and implemented by manufacturing companies. All the steps have to be recorded in detail, so that a complete history of every batch can be traced. GMP guidelines ensure that all operators are appropriately trained, and that all quality/safety controls are implemented.

Retroviral and lentiviral-based gene therapy products have to be manufactured following specific GMP guidelines. This is generally achieved by implementing streamlined production steps that minimize manipulation, composed of several purification/concentration steps, with quality controls for each batch. Particularly for lentivectors, the European Medicines Agency (EMA) provides guidelines on "the development and manufacture of lentiviral vectors" (CPMP/BWP/2458/03) and "guidance on the quality, preclinical and clinical aspects of gene transfer medicinal products" (CPMP/BWP/3088/99). These guidelines are quite broad and provide general directives regarding biological activities or levels of specific contaminants (especially from heterologous viral sequences/proteins such as VSV-G). These guidelines accept some residual degree of Gag–Pol transfer, and advise to set up assays for insertional mutagenesis and record integration sites in samples from genetically modified cells. Comprehensive in vitro and in vivo experiments assessing the specific characteristics of the therapeutic lentivectors and assays to detect replication-competent retroviruses (RCR) are required. However, these guidelines acknowledge that lentivector production by transient transfection results in less homogenous preparations than γ-retrovirus preparations using stable producer cell lines. The European Directorate for the Quality of Medicines & HealthCare (European Pharmacopoeia, which provides a reference work for specifications of pharmaceutical drugs) establishes a panel of tests to assure that the gene therapy product is safe and can be administered to human patients including sterility and adequate physico-chemical properties.

It is not the objective of GMP guidelines to dictate the specific production methods, and they assume that the most efficient manufacturing methods will be applied. For lentivector production, transient transfection is so far the most effective method, while stable producer/packaging cell lines have been successfully used in retroviral vector production. GMP guidelines have to be there in place to ensure the traceability and quality of the manufactured product.

5.3 Scaling-Up Lentivector Production for Clinical Application

As discussed above, one of the major steps for translating lentivector applications from animal experimental models to human application is scaling-up production.

5.3.1 Lentivector Titer and Particle Stability

For effective and realistic translation into human therapy, vector titers of around 10^6 infectious (transducing) particles have to be achieved, before further purification [6]. Therefore, a significant effort has been invested in improving lentivector performance to enhance transduction capacity and titers, as already discussed in Chapter 2. The lentivector transducing capacity will also depend on the envelope glycoprotein (EMV) used for pseudotyping. For clinical grade lentivector applications, the vesicular stomatitis virus G glycoprotein (VSV-G) has been used [3, 4, 7] because it confers high titers and particle stability. This is critical for scaling-up production to withstand the purification/concentration steps required for GMP production [6–8]. In a detailed characterization of a clinical grade VSV-G pseudotyped lentivector preparation for the correction of Wiscott-Aldrich syndrome, it was estimated that around 5×10^9 infectious particles were required for a correction of 5×10^6 CD34+ cells per patient (considering children with a 20 kg body mass).

Apart from VSV-G, other viral envelopes such as LCMV or baculovirus gp64 can also confer high particle stability, high titers and resistance to ultracentrifugation [9, 10]. This is in contrast to pseudotyping with γ-retrovirus ENVs, which usually provide lower titers and confer less stability [11]. An additional advantage of VSV-G pseudotypes is their resistance to freeze/thawing cycles. Therefore, it is not surprising that at least for ex vivo gene therapy approaches (those based on modification of bone marrow and reintroduction into the patient), VSV-G lentivector pseudotypes have been the vectors of choice [1, 3, 4, 6]. On the other hand, it is interesting to note that the first human gene therapy clinical trials used retroviral vectors pseudotyped with gibbon ape leukemia virus ENV [12] or amphotropic MLV ENV [13]. However, in this case supernatants from stable packaging/producer cells were directly incubated with target CD34+ progenitor cells.

5.3.2 Production Methods

Lentivectors generated for preclinical experimental models are widely produced by cotransfection of the vector transfer plasmid plus several plasmids encoding Gag–Pol, rev and an envelope glycoprotein for pseudotyping as discussed in Chapter 2 [3, 4, 6, 7]. Lentivectors have been classically produced by transfecting clones of human embryonic kidney cells (HEK) 293. This cell line was obtained by transformation with adenoviral early region 1 genes E1A and E1B [14]. From this original cell line, two additional clones were obtained to improve transfectability and protein expression. The most commonly used modified HEK cells are 293T cells, which constitutively express the simian virus 40 large T antigen [15]; an alternative are 293 EBNA-1 cells, which constitutively express the Epstein-Barr virus nuclear antigen 1 [16]. The advantages of these modified 293 cell lines are that they allow plasmid replication/episomal persistence (as long as the plasmids contain their corresponding SV40 or EBNA-1 replication origins) and also enhance the transcriptional activity of cellular and viral eukaryotic promoters. In addition, 293-based cell lines can be adapted to grow in suspension and in serum-free medium [17, 18]. These characteristics make 293-based cell lines attractive for large scale manufacturing processes required for protein expression/purification and lentivector production [18, 19]. From a biosafety point of view, 293 cells would be a better choice because they lack certain contaminants (DNA and protein) with oncogenic potential such as SV40 T antigen or EBNA-1. The first clinical grade lentivector was generated in 293 cells [7], and clones adapted to grow in suspension in serum-free medium have been successfully used for production of recombinant proteins and viral vectors, including lentivectors [17, 18, 20, 21]. However, 293T and 293 EBNA cells are more efficient producers, and grow much faster than 293 cells [6]. As a matter of fact, 293T and 293 EBNA cells have already been used to produce clinical grade biologicals, including lentivectors [6, 18, 19].

The two main differences between clinical grade lentivectors and those used for preclinical studies are the purity degree and traceability. To obtain the sufficient initial yield to provide clinically applicable vectors, large volumes of starting material (supernatants from producer cells) are required. In a detailed characterization of a clinical grade lentivector preparation [6], the authors estimated that 50 liters of lentivector-containing supernatant were required to treat five patients. Therefore, producer cells should be susceptible to large-scale growth in bioreactors or cell factories, easily applicable for GMP guidelines [6, 8]. So far, all derivatives of HEK-293 cells have been suitable for large-scale growth either as monolayer's in cell factory stacks with foetal calf serum (FCS) [6] or in suspension in the absence of FCS [17–19]. Interestingly, transient transfection has been the method of choice in all published large-scale lentivector-manufacturing methods. Calcium phosphate or polyethylenimine (PEI) was used for transfection due to its low cost [3, 4, 6].

Until recently, it was believed that long-term viral vector production from stable producer cells was the most efficient method to reproducibly obtain large volumes of vector preparation, and most importantly, for ease to follow GMP production guidelines. The engineering of stable producer lines would reduce the risk of DNA recombination between the cotransfected plasmids, and thus, reduce the potential source of replication-competent lentivectors [22]. Low passage GMP clinical grade cells could therefore be stored, and regrown for vector production. Stable producer cells have been successfully developed and used for large-scale production of γ-retroviral vectors. Different producer cells are commercially available with a range of pseudotyped retroviral envelopes. This is the case of the 3T3 PG13 producer/packaging cell line, which reproducibly generates vector titers around 10^6 transducing particles/ml [23, 24]. These vector preparations can be used without further concentration or purification [12]. Other producer/packaging retrovirus cell lines have also been used for production clinical grade vector production, such as 3T3-based ΨCrip. In this case, retroviruses are pseudotyped with amphotropic MLV ENV, leading to titers of at least 5×10^5 transducing particles per ml [13, 25]. In sharp contrast, large-scale clinical grade lentivectors have been produced in GMP conditions, but applying transient transfection instead of using stable producer/packaging cell lines.

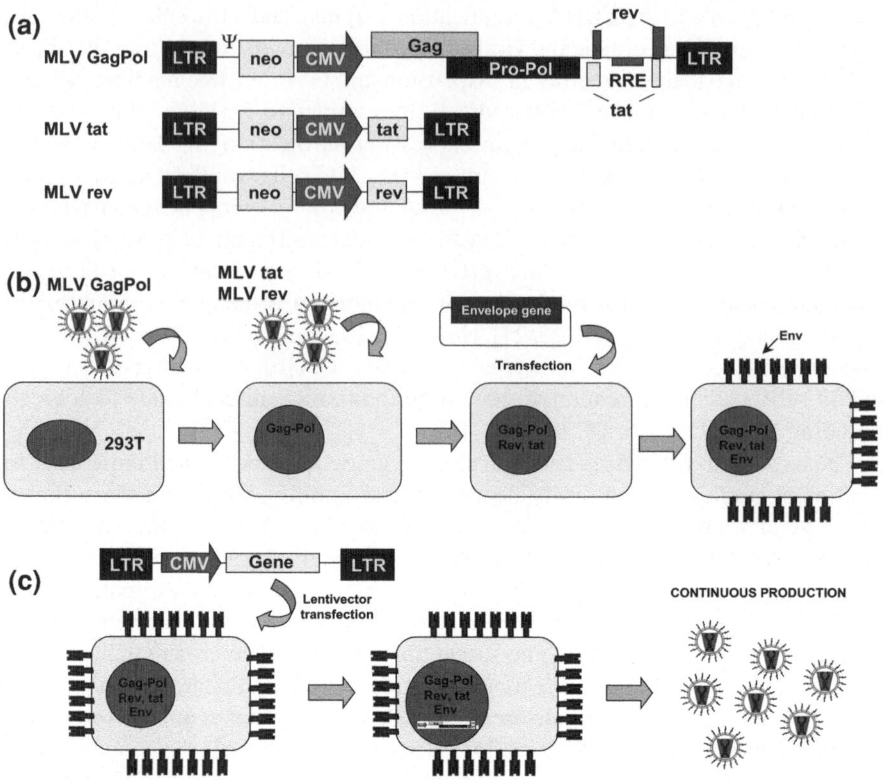

Fig. 5.1 Engineering of stable producing/packaging cell lines leading to continuous lentivector production. This figure depicts the necessary steps to generate long-term stable lentivector producing/packaging cell lines based on published protocols [27, 31]. (**a**) HIV-1 Gag–Pol, tat (if necessary) and rev are delivered within an MLV retrovirus vector. These vectors are produced following standard procedures and pseudotyped with VSV-G. Please note that in these constructs, the retroviral packaging signal is present. In addition, Gag–Pol, rev and tat are expressed from an internal CMV promoter, rather than by the MLV LTR. (**b**) The steps to generate a master 293T-based packaging/producer cell line are shown. Firstly, Gag–Pol is introduced after retroviral transduction (MLV GagPol) and clones are selected according to Gag–Pol expression and p24 and RT activities in the cell medium. Then, the most efficient, stable clone is selected and Tat-rev are introduced by retroviral transduction (MLV tat, MLV rev). Following cloning, cells are transfected with an expression plasmid for an envelope glycoprotein of interest, and cells are further cloned according to their capability of producing high titers of pseudotyped lentivectors. This cell line is then banked and stored. Note that this packaging cell line expresses the viral glycoprotein on the cell surface (trimeric envelope, in blue). (**c**) To produce a lentivector batch, cells from B are transfected with a lentivector plasmid of interest, and lentivectors either collected after transfection, or a clone is again selected which produces high lentivector titers. These cells produce lentivectors up to several months

The engineering of stable producer/packaging cell lines for HIV-1-based lentivectors is a technical challenge. Firstly, VSV-G cannot be constitutively expressed because of its toxicity. Secondly, constitutive expression of HIV-1 Gag–Pol has also been found to be particularly toxic. Huge effort has been directed to solve these two main issues. One of the first successful stable cell lines was based on the inducible expression of VSV-G, Rev, and Gag–Pol in 293 cells based on the Tet-off system [22]. In the presence of doxycycline, low amounts of VSV-G and Gag–Pol products were expressed. These cells were used to produce third-generation lentivectors, and they could be kept in culture in the absence of doxycycline for 2 weeks before cell death. Production of around 1–20 HeLa-transducing units per cell per day could be achieved for 1 week [22, 26]. Both strategies were based on the Tet-off system, and although they showed that titers obtained with these producer cell lines were comparable to transient transfection, the constant administration of doxycycline to repress VSV-G and Gag–Pol expression would likely hamper their adaptation to large-scale GMP production. In 2003, a breakthrough on the engineering of lentivector stable producer cell lines was achieved [27]. It was always believed that constitutive expression of Gag–Pol (with the HIV protease), Tat, and Rev significantly contributed to the difficulties of selecting effective packaging clones [28]. In fact, this may be true after plasmid transfection followed by antibiotic selection. Curiously, stable 293T cell clones constitutively expressing Gag–Pol, Rev, and Tat were obtained in a straightforward way. Instead of plasmid transfection, a codon-optimized Gag–Pol construct, Tat, and Rev were integrated into 293T cells using MLV retroviral vectors (Fig. 5.1). Several stable cell lines were engineered which pseudotyped lentivectors with a range of γ-retroviral and alphaviral ENVs [11, 27, 29]. These stable packaging cells produced up to 10^7 particles per ml for at least 3 months in culture [27], making them perfect candidates for large scale production under GMP guidelines. However, these cells were not obtained under GMP conditions, and have not been applied for human therapy yet. Other producer cell lines based on transfection, and regulated expression systems have been published, although with similar efficiencies to other systems [30]. Recently, a novel producer cell line was constructed again using γ-retroviral vectors to introduce codon-optimized HIV components, together with a Tet-off regulatable system for Rev and VSV-G expression [31]. Importantly, while this strategy is similar to a previously reported one [27], these producer cell lines were generated from traceable 293T cells (Fig. 5.1). Production of self-inactivating lentivectors was successfully scaled up in bioreactors leading to the production of 20l supernatants with vector titers higher than 10^7 transducing particles per ml [31]. This last system demonstrates that the application of stable producer cell lines for clinical grade lentivectors is closer than ever before.

5.4 Purity of Clinical Grade Lentivectors

Preclinical lentivectors used in experimental models are produced by small-scale transient transfection. Supernatants containing lentivector particles are collected for several days (usually up to 1 week) always in the presence of foetal calf serum, and filtered to remove large contaminants such as cellular debris. After that, they are either used directly or concentrated by ultracentrifugation, and sometimes slow-speed centrifugation depending on the particle stability. These concentration

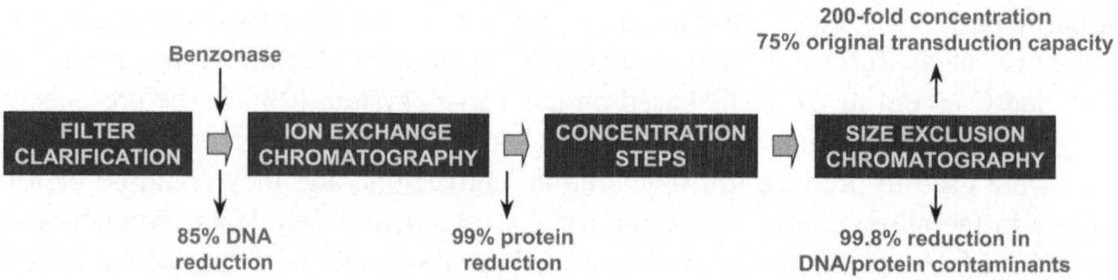

Fig. 5.2 Large-scale production/manufacturing process for a clinical grade lentiviral vector. This scheme summarizes the main steps undertaken to obtain a clinical grade lentiviral vector as described by Merten and collaborators [6]. The key major steps are shown within filled boxes. The benzonase step is also indicated (top left arrow). The reduction of contaminating DNA and protein in each step is indicated below in red lettering. On the upper right part of the figure, a summary on the properties of the final clinical grade vector preparation is also indicated. Other steps have been removed from the scheme for simplification purposes. There are other published procedures for the manufacturing of large-scale lentivectors which are similar, but use other chromatographic steps such as heparin affinity chromatography [18, 19]

steps also concentrate other small-sized contaminants such as protein clumps, and cell-derived vesicles. Therefore, preclinical lentivector preparations are inherently unsuitable for use in human therapy.

The lack of consistency in the way lentivector preparations have been produced has caused confusion about their effects in experimental models of disease. As an example, it is still unclear whether lentivector particles themselves are immunogenic or not. There are reports suggesting that lentivector transduction can activate dendritic cells (DCs), while other reports suggest that this is caused by contaminants in lentivector preparations [32–35]. Contaminants in lentivector preparations pose a real danger if used in gene therapy applications to correct genetic diseases [36]. In fact, purer lentivector preparations have lower inflammatory capacities [37]. Lentivector preparations can be purified by sucrose gradient ultracentrifugation, or production in the absence of FCS [37, 38]. For some experimental settings, concentration through a sucrose cushion is enough to decrease their capacities to induce DC maturation ex vivo [33]. However, it has to be taken into account that lentivectors are virus-like particles and they are still recognized by pathogen recognition receptors [32].

The process of scaling-up production followed by concentration increases the likelihood of contamination. To achieve clinical grade lentivectors, several purification steps have to be implemented (Fig. 5.2). There are at least three main sources of contamination; firstly, plasmid DNA from transfection; secondly, from the culture medium (presence/absence of FCS, antibiotics); and thirdly, contaminants of cellular origin. The source of DNA contamination comes from plasmid DNA used during transfection, or encapsulated plasmid DNA within VSV-G tubulovesicular structures, with the potential risk of DNA mobilization to target cells [39]. Therefore, a detection step of contaminating DNA has to be implemented during production.

To purify lentivectors up to clinical standards, several purification steps have to be streamlined to remove as much nucleic acids and protein contaminants as possible, without significantly affecting lentivector performance. Such a protocol is exemplified by Merten and collaborators, who published in detail the GMP production protocol of a clinical grade lentivector for the treatment of Wiscott–Aldrich syndrome (Fig. 5.2) [6]. In their protocol, they identified specific contaminants which would pose a definite biosafety risk. Totally, 293T cells were used for production, from which adenoviral E1A, E1B genes, SV40 T antigen, and other undefined cellular contaminants could be expected. Large-scale transient transfection was performed, and therefore, a potential source of plasmid DNA contaminant. During the purification process, these contaminants were monitored by total protein content and host protein levels by spectrophotometric techniques and ELISA. Total DNA content was also monitored by spectrophotometry, and specific quantitative-PCR (Q-PCR) tests were designed for detection of E1A, SV40 T, and VSV-G DNA sequences. As mentioned before, poorly purified lentivector preparations can transfer DNA plasmids to target cells [39], so their transfer was also assessed by Q-PCR in target cells after transduction and culture during six passages. Their manufacturing protocol was separated in various purification steps, starting with clarification and filter of supernatants, and an overnight treatment with benzonase to remove DNA contaminants (Fig. 5.2) [40, 41]. The benzonase step seemed to be essential because it is unclear whether other purification techniques such as chromatography completely remove contaminating DNA [42]. Just the benzonase step eliminated about 85% of DNA content, and a further ion-exchange chromatography removed approximately 99% of protein contamintants [6]. These purification steps were coupled to further vector concentration, and purification by size exclusion chromatography (Fig. 5.2) [6]. Chromatographic purification techniques have proven to be reliable for the processing of large-scale vector production, leading to the removal of most contaminants and vector concentration [6, 8, 17–19, 42]. Merten and collaborators achieved a reduction in total DNA and protein content of around 99.8%, resulting in a concentration factor of 200 (final volume 240 ml), with lentivector titers between 10^7–10^9 transducing units per ml.

5.5 Biosafety

Biosafety is a key factor nowadays in human gene therapy, especially after the first deaths associated with early gene therapy trials. In 1999, a patient died of high fever and blood clots as a result of serious failing in the implementation of a gene therapy protocol using adenovirus vectors [43–45]. This trial was aimed at the correction of partial ornithine transcarbamylase deficiency [46, 47]. In 2006, one patient died after correction of chronic granulomatous disease (CGD) using γ-retroviral vectors, although his death might have been unrelated to the gene therapy itself [48, 49]. In 2007, a patient died after gene therapy for the treatment of rheumatoid arthritis using an adeno-associated virus vector delivering a decoy TNF receptor fused to a

Fc by local injection in the joint [50, 51]. This patient died of *Histoplasma* infection, although it is still unclear whether the gene therapy treatment played a role in this death [50]. In the most successful human clinical trials so far for the correction of SCID-X1, several deaths also occurred after leukemia caused by insertional mutagenesis [52]. In addition, replication-competent retrovirus vectors arising from a retrovirus vector preparation used for genetic modification of CD34+ hematopoietic stem cells resulted in leukemia and death of several nonhuman primates [53]. This RCR appeared after recombination in the producer cells, highlighting the need for the establishment of RCR detection tests [44, 53, 54].

5.5.1 Insertional Mutagenesis

The hallmark of retroviral and lentiviral vectors is their stable integration in the host cell chromosomes. However, this integration can upregulate protooncogenes or inactivate tumor-suppressor genes, a process called insertional mutagenesis [55, 56]. The occurrence of leukemia in some of the children treated for correction of SCID-X1 using γ-retrovirus vectors has highlighted the necessity for genotoxicity assays [57]. However, there are still some doubts about the information that can be derived from these assays since genotoxicity could be linked to other causes apart from insertional mutagenesis [58, 59]. Since the first human gene therapy trials, several assays, and experimental systems have been set up to test for genotoxicity, and a few examples are discussed in detail below.

5.5.1.1 Cell-Based Assays

In vitro cell-based mutagenesis assays could be routinely implemented to test the insertional mutagenesis potential of clinical grade lentivector batches. These assays would be advantageous due to their low cost and potentially high reproducibility. One of such assays was set up to quantify the mutagenic potential of retroviral and lentiviral vectors by analyzing their capacity to confer growth-factor independency to cell lines [55]. Thus, the IL-3-dependent immortalized mouse cell line BAF3 can be rendered IL-3-independent through retroviral integration [55, 60]. In this assay, BAF3 cells are transduced with the vectors of interest and IL-3 is then removed from the medium. This system allowed the isolation of an IL-3-independent colony. To further increase its sensitivity, the bcl-2-overexpressing BAF3 mutant Bcl15 was used to favor the rescue of retrolentivector-generated mutants. Very interestingly, only one type of mutants was selected in this assay, after insertional upregulation of the growth hormone receptor (GHR). This upregulation was caused by run-off transcription from the vector U3 followed by splicing which resulted in mRNAs encoding the GHR. γ-retroviral vectors also generated mutants but by insertion in other loci, one of them IL-3 itself and several other common integration sites (CIS). Further characterization of these mutants showed that the internal promoter present in the lentivector itself enhanced the LTR transcriptional activity [61]. Deletion of the U3 enhancers (self-inactivating lentivector) greatly diminished GHR upregulation [62].

The application of the previous assay is straightforward and reproducible. In addition, it sheds light onto the mechanisms used by retroviral and lentiviral vectors to generate cell growth-independent mutants, which may correlate with their genotoxic potential. However, the significance of this assay for translation into human gene therapy is unclear. Other cell-based assays have attempted to re-create gene therapy protocols in animal models but focused on detection of cell transformation. In this regard, lentivector transduction of mouse hematopoietic cells followed by clonal dilution demonstrated that a major transforming determinant was a strong internal LTR promoter even in the context of a self-inactivating lentivector [63], in agreement with results from the BAF3-based assay [61]. However, it is worth noting that SIN lentivectors showed a reduced mutagenic potential compared to nonSIN lentivectors, again in agreement with the BAF-based system [62]. However, unlike the BAF3-based assay, integrations occurred nearby the CIS Evi1. Both assays also showed that lentivectors were generally less genotoxic than retroviral vectors, exhibiting differential insertional patterns. A clinical grade lentivector for the treatment of Wiscott-Aldrich syndrome, containing an endogenous cellular promoter did not exhibit detectable mutagenic capacities [64]. Basically, the same results were obtained for a lentivector containing the β-globin locus control region in an in vitro immortalization assay using lentivector-transduced mouse hematopoietic precursors, and in general when cellular promoters are used [63, 65, 66]. The inclusion of insulator sequences may have also contributed to the reduction in genotoxicity [65]. The combined results from the two cell-based systems show that SIN vectors are significantly less genotoxic, that the nature of the internal promoter plays a key role, and that lentivectors exert different molecular mutagenic mechanisms depending on the type of assay.

5.5.1.2 In Vivo Assays

Cell-based assays have the advantage that they are reproducible and relatively simple to set up. However, their results are somewhat difficult to translate to a complex organism. To overcome this problem, insertional mutagenesis has been studied in several in vivo assays. In some of these assays, the retroviruses themselves contained oncogenes such as Sox4, and their attention was focused on the identification of oncogene upregulation which could cooperate in the induction of leukemia by insertional mutagenesis [67]. Retrovirus expression of the large SV40 tumor antigen was also used to increase the sensitivity of detection of insertional mutagenesis in mice transplanted with transduced hematopoietic precursors. Mice developed leukemic cells with fast kinetics, and insertion sites occurred close to genes modulating cell division and apoptosis [68]. Overall, these results clearly showed that insertion in cells that may be prone to transformation could increase the risk of in vivo tumorigenesis.

Finally, both in vivo and in vitro assays allow the identification of CIS by γ-retroviral and lentiviral vectors, and the characterization of important mutagenic mechanisms which could be therapeutically relevant [61, 62, 69, 70].

5.5.2 Detection of Replication Competent Retroviruses/Lentiviruses (RCR/RCL)

Regulatory agencies require the implementation of tests for detection of replication competent retrovirus/lentivirus in preparations of clinical grade vectors. RCR/RCLs arise from recombination between the transfected plasmid molecules, and even from retroviral-like DNA sequences in producer cells [54]. This is particularly important in large-scale vector production, which increases the probability of recombination events. The need for the implementation of sensitive detection assays for RCRs and RCLs was highlighted after some deaths of nonhuman primates in experimental models of gene therapy as a consequence of leukemia after RCR integration [53]. Additionally, transient transfection is thought to be riskier than production using stable producer/packaging cells [26, 27, 54]. To prevent the appearance of RCLs, the vector genome has been fragmented into at least three plasmids. This requires the occurrence of multiple recombination events to generate RCLs. Additionally, other safety measures such as lack of packaging signals other than in the transfer vector, and the engineering of self-inactivating vectors make sure that in the unlikely event of multiple recombinations, only nonviable, nonpathogenic recombinant viruses would appear.

A key point in RCR and RCL detection assays is the design of appropriate positive controls [44], which will provide an estimate of the sensitivity of the assay, which should be usually in the order of 1 RCR/RCL per 10^8–10^9 particles [71]. Merten and collaborators used an attenuated strain of HIV-1 without accessory genes, pseudotyped with VSV-G or HIV Env as a positive control. This assay could detect as little as 10 femtograms of infectious p24 as measured by 50% tissue culture infectious dose (TCID50). In addition, the authors demonstrated by "spiking" the positive control that the actual lentivector preparations did not inhibit the detection assay.

No evidence for vector-derived RLC was observed in any of the human clinical trials with lentivectors or γ-retroviral vectors [3, 4, 12, 13, 49], especially in the first lentivector trial for the treatment of HIV [2]. Due to the presence of endogenous HIV, there were fears of recombinant lentivectors with "novel" properties due to recombination between the vector and replicating wild-type viruses.

5.6 Final Considerations and Conclusions

In the previous sections, lentivector production, contaminants, and biosafety considerations have been discussed to some detail. Before human use, clinical grade lentivectors have to undergo a series of extensive quality controls. Many of these tests are directed to quantify contaminants arising from production, as discussed before. In addition, before administration in human patients, microbiological tests (including endotoxin levels) have to be carried out to make sure that the final product can be administered in vivo without causing harm.

To conclude, it has been demonstrated that GMP clinical grade lentivectors can be produced by transient transfection in 293-based cell lines. In contrast, the development of stable producer/packaging cells lines is taking longer than anticipated, although there are at least two promising strategies that could be implemented for GMP production [27, 31]. The development of these packaging cells may further diminish the costs of production and make gene therapy with lentivectors "routine" therapeutic treatments.

Acknowledgments David Escors is funded by an Arthritis Research UK Career Development Fellowship (18433). Holly Stephenson is funded by the Biomedical Research Centre, Institute of Child Health, UCL. Karine Breckpot is funded by the Fund for Scientific Research-Flandes. The Oxford Structural Genomics Consortium is a registered UK charity (number 1097737) that receives funds from the Canadian Institutes of Health Research, The Canadian Foundation for Innovation, Genome Canada through the Ontario Genomics Institute, GlaxoSmithKline, Karolinska Institutet, the Knut and Alice Wallenberg Foundations, the Ontario Innovation Trust, the Ontario Ministry for Research and Innovation, Merck & Co., Inc., the Novartis Research Foundation, the Swedish Foundation for Strategic Research and the Wellcome Trust.

References

1. DiGiusto DL, Krishnan A, Li L, Li H, Li S, Rao A, Mi S, Yam P, Stinson S, Kalos M, Alvarnas J, Lacey SF, Yee JK, Li M, Couture L, Hsu D, Forman SJ, Rossi JJ, Zaia JA (2010) RNA-based gene therapy for HIV with lentiviral vector-modified CD34(+) cells in patients undergoing transplantation for AIDS-related lymphoma. Sci Transl Med 2(36):36ra43
2. Levine BL, Humeau LM, Boyer J, MacGregor RR, Rebello T, Lu X, Binder GK, Slepushkin V, Lemiale F, Mascola JR, Bushman FD, Dropulic B, June CH (2006) Gene transfer in humans using a conditionally replicating lentiviral vector. Proc Natl Acad Sci U S A 103(46): 17372–17377
3. Cartier N, Hacein-Bey-Abina S, Bartholomae CC, Veres G, Schmidt M, Kutschera I, Vidaud M, Abel U, Dal-Cortivo L, Caccavelli L, Mahlaoui N, Kiermer V, Mittelstaedt D, Bellesme C, Lahlou N, Lefrere F, Blanche S, Audit M, Payen E, Leboulch P, l'Homme B, Bougneres P, Von Kalle C, Fischer A, Cavazzana-Calvo M, Aubourg P (2009) Hematopoietic stem cell gene therapy with a lentiviral vector in X-linked adrenoleukodystrophy. Science 326(5954):818–823
4. Cavazzana-Calvo M, Payen E, Negre O, Wang G, Hehir K, Fusil F, Down J, Denaro M, Brady T, Westerman K, Cavallesco R, Gillet-Legrand B, Caccavelli L, Sgarra R, Maouche-Chretien L, Bernaudin F, Girot R, Dorazio R, Mulder GJ, Polack A, Bank A, Soulier J, Larghero J, Kabbara N, Dalle B, Gourmel B, Socie G, Chretien S, Cartier N, Aubourg P, Fischer A, Cornetta K, Galacteros F, Beuzard Y, Gluckman E, Bushman F, Hacein-Bey-Abina S, Leboulch P (2010) Transfusion independence and HMGA2 activation after gene therapy of human beta-thalassaemia; 1476-4687 (Electronic) 0028-0836 (Linking); Sep 16, 2010; pp 318–322
5. Boztug K, Schmidt M, Schwarzer A, Banerjee PP, Diez IA, Dewey RA, Bohm M, Nowrouzi A, Ball CR, Glimm H, Naundorf S, Kuhlcke K, Blasczyk R, Kondratenko I, Marodi L, Orange JS,

von Kalle C, Klein C (2010) Stem-cell gene therapy for the Wiskott-Aldrich syndrome. The New England journal of medicine 363(20):1918–1927
6. Merten OW, Charrier S, Laroudie N, Fauchille S, Dugue C, Jenny C, Audit M, Zanta-Boussif MA, Chautard H, Radrizzani M, Vallanti G, Naldini L, Noguiez-Hellin P, Galy A (2011) Large-scale manufacture and characterization of a lentiviral vector produced for clinical ex vivo gene therapy application. Hum Gene Ther 22(3):343–356
7. Lu X, Humeau L, Slepushkin V, Binder G, Yu Q, Slepushkina T, Chen Z, Merling R, Davis B, Chang YN, Dropulic B (2004) Safe two-plasmid production for the first clinical lentivirus vector that achieves >99% transduction in primary cells using a one-step protocol. J Gene Med 6(9):963–973
8. Slepushkin V, Chang N, Cohen R, Gan Y, Jiang B, Deausen E, Berlinger D, Binder G, Andre K, Humeau L, Dropulic B (2003) Large-scale purification of a lentiviral vector by size exclusion chromatography or mustang Q ion Exchange Capsule. Bioprocess J 2:89–95
9. Beyer WR, Westphal M, Ostertag W, von Laer D (2002) Oncoretrovirus and lentivirus vectors pseudotyped with lymphocytic choriomeningitis virus glycoprotein: generation, concentration, and broad host range. J Virol 76(3):1488–1495
10. Schauber CA, Tuerk MJ, Pacheco CD, Escarpe PA, Veres G (2004) Lentiviral vectors pseudotyped with baculovirus gp64 efficiently transduce mouse cells in vivo and show tropism restriction against hematopoietic cell types in vitro. Gene Ther 11(3):266–275
11. Strang BL, Ikeda Y, Cosset FL, Collins MK, Takeuchi Y (2004) Characterization of HIV-1 vectors with gammaretrovirus envelope glycoproteins produced from stable packaging cells. Gene Ther 11(7):591–598
12. Gaspar HB, Parsley KL, Howe S, King D, Gilmour KC, Sinclair J, Brouns G, Schmidt M, Von Kalle C, Barington T, Jakobsen MA, Christensen HO, Al Ghonaium A, White HN, Smith JL, Levinsky RJ, Ali RR, Kinnon C, Thrasher AJ (2004) Gene therapy of X-linked severe combined immunodeficiency by use of a pseudotyped gammaretroviral vector. Lancet 364(9452):2181–2187
13. Cavazzana-Calvo M, Hacein-Bey S, de Saint Basile G, Gross F, Yvon E, Nusbaum P, Selz F, Hue C, Certain S, Casanova JL, Bousso P, Deist FL, Fischer A (2000) Gene therapy of human severe combined immunodeficiency (SCID)-X1 disease. Science 288(5466):669–672
14. Graham FL, Smiley J, Russell WC, Nairn R (1977) Characteristics of a human cell line transformed by DNA from human adenovirus type 5. J Gen Virol 36(1):59–74
15. DuBridge RB, Tang P, Hsia HC, Leong PM, Miller JH, Calos MP (1987) Analysis of mutation in human cells by using an Epstein-Barr virus shuttle system. Mol Cell Biol 7(1): 379–387
16. Young JM, Cheadle C, Foulke JS Jr, Drohan WN, Sarver N (1988) Utilization of an Epstein-Barr virus replicon as a eukaryotic expression vector. Gene 62(2):171–185
17. Ansorge S, Lanthier S, Transfiguracion J, Durocher Y, Henry O, Kamen A (2009) Development of a scalable process for high-yield lentiviral vector production by transient transfection of HEK293 suspension cultures. J Gene Med 11(10):868–876
18. Segura MM, Garnier A, Durocher Y, Ansorge S, Kamen A (2010) New protocol for lentiviral vector mass production. Methods Mol Biol 614:39–52
19. Segura MM, Garnier A, Durocher Y, Coelho H, Kamen A (2007) Production of lentiviral vectors by large-scale transient transfection of suspension cultures and affinity chromatography purification. Biotechnol Bioeng 98(4):789–799
20. Cote J, Bourget L, Garnier A, Kamen A (1997) Study of adenovirus production in serum-free 293SF suspension culture by GFP-expression monitoring. Biotechnol Prog 13(6):709–714
21. Cote J, Garnier A, Massie B, Kamen A (1998) Serum-free production of recombinant proteins and adenoviral vectors by 293SF-3F6 cells. Biotechnol Bioeng 59(5):567–575
22. Klages N, Zufferey R, Trono D (2000) A stable system for the high-titer production of multiply attenuated lentiviral vectors. Mol Ther 2(2):170–176
23. Miller AD, Garcia JV, von Suhr N, Lynch CM, Wilson C, Eiden MV (1991) Construction and properties of retrovirus packaging cells based on gibbon ape leukemia virus. J Virol 65(5):2220–2224

24. Loew R, Meyer Y, Kuehlcke K, Gama-Norton L, Wirth D, Hauser H, Stein S, Grez M, Thornhill S, Thrasher A, Baum C, Schambach A (2010) A new PG13-based packaging cell line for stable production of clinical-grade self-inactivating gamma-retroviral vectors using targeted integration. Gene Ther 17(2):272–280
25. Danos O, Mulligan RC (1988) Safe and efficient generation of recombinant retroviruses with amphotropic and ecotropic host ranges. Proc Natl Acad Sci U S A 85(17):6460–6464
26. Farson D, Witt R, McGuinness R, Dull T, Kelly M, Song J, Radeke R, Bukovsky A, Consiglio A, Naldini L (2001) A new-generation stable inducible packaging cell line for lentiviral vectors. Hum Gene Ther 12(8):981–997
27. Ikeda Y, Takeuchi Y, Martin F, Cosset FL, Mitrophanous K, Collins M (2003) Continuous high-titer HIV-1 vector production. Nat Biotechnol 21(5):569–572
28. Kaplan AH, Swanstrom R (1991) The HIV-1 gag precursor is processed via two pathways: implications for cytotoxicity. Biomed Biochim Acta 50(4–6):647–653
29. Strang BL, Takeuchi Y, Relander T, Richter J, Bailey R, Sanders DA, Collins MK, Ikeda Y (2005) Human immunodeficiency virus type 1 vectors with alphavirus envelope glycoproteins produced from stable packaging cells. J Virol 79(3):1765–1771
30. Ni Y, Sun S, Oparaocha I, Humeau L, Davis B, Cohen R, Binder G, Chang YN, Slepushkin V, Dropulic B (2005) Generation of a packaging cell line for prolonged large-scale production of high-titer HIV-1-based lentiviral vector. J Gene Med 7(6):818–834
31. Throm RE, Ouma AA, Zhou S, Chandrasekaran A, Lockey T, Greene M, De Ravin SS, Moayeri M, Malech HL, Sorrentino BP, Gray JT (2009) Efficient construction of producer cell lines for a SIN lentiviral vector for SCID-X1 gene therapy by concatemeric array transfection. Blood 113(21):5104–5110
32. Breckpot K, Escors D, Arce F, Lopes L, Karwacz K, Van Lint S, Keyaerts M, Collins M (2010) HIV-1 lentiviral vector immunogenicity is mediated by Toll-like receptor 3 (TLR3) and TLR7. J Virol 84:5627–5636
33. Escors D, Lopes L, Lin R, Hiscott J, Akira S, Davis RJ, Collins MK (2008) Targeting dendritic cell signalling to regulate the response to immunisation. Blood 111(6):3050–3061
34. Goold HD, Escors D, Conlan TJ, Chakraverty R, Bennett CL (2011) Conventional DC are required for the activation of helper-dependent CD8 T cell responses after cutaneous vaccination with lentiviral vectors. J Immunol 186(8):4565–4572
35. Reiser J (2000) Production and concentration of pseudotyped HIV-1-based gene transfer vectors. Gene Ther 7(11):910–913
36. Tuschong L, Soenen SL, Blaese RM, Candotti F, Muul LM (2002) Immune response to fetal calf serum by two adenosine deaminase-deficient patients after T cell gene therapy. Hum Gene Ther 13(13):1605–1610
37. Baekelandt V, Eggermont K, Michiels M, Nuttin B, Debyser Z (2003) Optimized lentiviral vector production and purification procedure prevents immune response after transduction of mouse brain. Gene Ther 10(23):1933–1940
38. Escors D, Capiscol C, Enjuanes L (2004) Immunopurification applied to the study of virus protein composition and encapsidation. J Virol Methods 119(2):57–64
39. Pichlmair A, Diebold SS, Gschmeissner S, Takeuchi Y, Ikeda Y, Collins MK, Reis e Sousa C (2007) Tubulovesicular structures within vesicular stomatitis virus G protein-pseudotyped lentiviral vector preparations carry DNA and stimulate antiviral responses via Toll-like receptor 9. J Virol 81(2):539–547
40. Sastry L, Xu Y, Cooper R, Pollok K, Cornetta K (2004) Evaluation of plasmid DNA removal from lentiviral vectors by benzonase treatment. Hum Gene Ther 15(2):221–226
41. Zufferey R (2002) Production of lentiviral vectors. Curr Top Microbiol Immunol 261:107–121
42. Yamada K, McCarty DM, Madden VJ, Walsh CE (2003) Lentivirus vector purification using anion exchange HPLC leads to improved gene transfer. BioTechniques 34(5):1074–1078, 1080
43. Lehrman S (1999) Virus treatment questioned after gene therapy death. Nature 401(6753): 517–518
44. Manilla P, Rebello T, Afable C, Lu X, Slepushkin V, Humeau LM, Schonely K, Ni Y, Binder GK, Levine BL, MacGregor RR, June CH, Dropulic B (2005) Regulatory

considerations for novel gene therapy products: a review of the process leading to the first clinical lentiviral vector. Hum Gene Ther 16(1):17–25
45. Marshall E (1999) Gene therapy death prompts review of adenovirus vector. Science 286(5448):2244–2245
46. Raper SE, Yudkoff M, Chirmule N, Gao GP, Nunes F, Haskal ZJ, Furth EE, Propert KJ, Robinson MB, Magosin S, Simoes H, Speicher L, Hughes J, Tazelaar J, Wivel NA, Wilson JM, Batshaw ML (2002) A pilot study of in vivo liver-directed gene transfer with an adenoviral vector in partial ornithine transcarbamylase deficiency. Hum Gene Ther 13(1):163–175
47. Raper SE, Chirmule N, Lee FS, Wivel NA, Bagg A, Gao GP, Wilson JM, Batshaw ML (2003) Fatal systemic inflammatory response syndrome in a ornithine transcarbamylase deficient patient following adenoviral gene transfer. Mol Genet Metab 80(1–2):148–158
48. European Society of Gene Therapy (ESGT) (2006) One of three successfully treated CGD patients in a Swiss–German gene therapy trial died due to his underlying disease: a position statement from the European Society of Gene Therapy (ESGT). J Gene Med 8(12):1435
49. Ott MG, Schmidt M, Schwarzwaelder K, Stein S, Siler U, Koehl U, Glimm H, Kuhlcke K, Schilz A, Kunkel H, Naundorf S, Brinkmann A, Deichmann A, Fischer M, Ball C, Pilz I, Dunbar C, Du Y, Jenkins NA, Copeland NG, Luthi U, Hassan M, Thrasher AJ, Hoelzer D, von Kalle C, Seger R, Grez M (2006) Correction of X-linked chronic granulomatous disease by gene therapy, augmented by insertional activation of MDS1-EVI1, PRDM16 or SETBP1. Nat Med 12(4):401–409
50. Evans CH, Ghivizzani SC, Robbins PD (2008) Arthritis gene therapy's first death. Arthritis Res Ther 10(3):110
51. Kaiser J (2007) Clinical research. Death prompts a review of gene therapy vector. Science 317(5838):580
52. Thrasher AJ, Gaspar HB, Baum C, Modlich U, Schambach A, Candotti F, Otsu M, Sorrentino B, Scobie L, Cameron E, Blyth K, Neil J, Abina SH, Cavazzana-Calvo M, Fischer A (2006) Gene therapy: X-SCID transgene leukaemogenicity. Nature 443(7109):E5–6, discussion E6–7
53. Donahue RE, Kessler SW, Bodine D, McDonagh K, Dunbar C, Goodman S, Agricola B, Byrne E, Raffeld M, Moen R et al (1992) Helper virus induced T cell lymphoma in nonhuman primates after retroviral mediated gene transfer. J Exp Med 176(4):1125–1135
54. Sinn PL, Sauter SL, McCray PB Jr (2005) Gene therapy progress and prospects: development of improved lentiviral and retroviral vectors–design, biosafety, and production. Gene Ther 12(14):1089–1098
55. Stocking C, Loliger C, Kawai M, Suciu S, Gough N, Ostertag W (1988) Identification of genes involved in growth autonomy of hematopoietic cells by analysis of factor-independent mutants. Cell 53(6):869–879
56. Woods NB, Muessig A, Schmidt M, Flygare J, Olsson K, Salmon P, Trono D, von Kalle C, Karlsson S (2003) Lentiviral vector transduction of NOD/SCID repopulating cells results in multiple vector integrations per transduced cell: risk of insertional mutagenesis. Blood 101(4):1284–1289
57. Howe SJ, Mansour MR, Schwarzwaelder K, Bartholomae C, Hubank M, Kempski H, Brugman MH, Pike-Overzet K, Chatters SJ, de Ridder D, Gilmour KC, Adams S, Thornhill SI, Parsley KL, Staal FJ, Gale RE, Linch DC, Bayford J, Brown L, Quaye M, Kinnon C, Ancliff P, Webb DK, Schmidt M, von Kalle C, Gaspar HB, Thrasher AJ (2008) Insertional mutagenesis combined with acquired somatic mutations causes leukemogenesis following gene therapy of SCID-X1 patients. J Clin Invest 118(9):3143–3150
58. Maetzig T, Brugman MH, Bartels S, Heinz N, Kustikova OS, Modlich U, Li Z, Galla M, Schiedlmeier B, Schambach A, Baum C (2011) Polyclonal fluctuation of lentiviral vector-transduced and expanded murine hematopoietic stem cells. Blood 117(11):3053–3064
59. Ginn SL, Liao SH, Dane AP, Hu M, Hyman J, Finnie JW, Zheng M, Cavazzana-Calvo M, Alexander SI, Thrasher AJ, Alexander IE (2010) Lymphomagenesis in SCID-X1 mice following lentivirus-mediated phenotype correction independent of insertional mutagenesis and gammac overexpression. Mol Ther 18(5):965–976

60. Thomas J, Leverrier Y, Marvel J (1998) Bcl-X is the major pleiotropic anti-apoptotic gene activated by retroviral insertion mutagenesis in an IL-3 dependent bone marrow derived cell line. Oncogene 16(11):1399–1408
61. Knight S, Bokhoven M, Collins M, Takeuchi Y (2010) Effect of the internal promoter on insertional gene activation by lentiviral vectors with an intact HIV long terminal repeat. J Virol 84(9):4856–4859
62. Bokhoven M, Stephen SL, Knight S, Gevers EF, Robinson IC, Takeuchi Y, Collins MK (2009) Insertional gene activation by lentiviral and gammaretroviral vectors. J Virol 83(1): 283–294
63. Modlich U, Bohne J, Schmidt M, von Kalle C, Knoss S, Schambach A, Baum C (2006) Cell-culture assays reveal the importance of retroviral vector design for insertional genotoxicity. Blood 108(8):2545–2553
64. Modlich U, Navarro S, Zychlinski D, Maetzig T, Knoess S, Brugman MH, Schambach A, Charrier S, Galy A, Thrasher AJ, Bueren J, Baum C (2009) Insertional transformation of hematopoietic cells by self-inactivating lentiviral and gammaretroviral vectors. Mol Ther 17(11):1919–1928
65. Arumugam PI, Higashimoto T, Urbinati F, Modlich U, Nestheide S, Xia P, Fox C, Corsinotti A, Baum C, Malik P (2009) Genotoxic potential of lineage-specific lentivirus vectors carrying the beta-globin locus control region. Mol Ther 17(11):1929–1937
66. Zychlinski D, Schambach A, Modlich U, Maetzig T, Meyer J, Grassman E, Mishra A, Baum C (2008) Physiological promoters reduce the genotoxic risk of integrating gene vectors. Mol Ther 16(4):718–725
67. Du Y, Spence SE, Jenkins NA, Copeland NG (2005) Cooperating cancer-gene identification through oncogenic-retrovirus-induced insertional mutagenesis. Blood 106(7):2498–2505
68. Li Z, Kustikova OS, Kamino K, Neumann T, Rhein M, Grassman E, Fehse B, Baum C (2007) Insertional mutagenesis by replication-deficient retroviral vectors encoding the large T oncogene. Ann N Y Acad Sci 1106:95–113
69. Kustikova OS, Schiedlmeier B, Brugman MH, Stahlhut M, Bartels S, Li Z, Baum C (2009) Cell-intrinsic and vector-related properties cooperate to determine the incidence and consequences of insertional mutagenesis. Mol Ther 17(9):1537–1547
70. Maruggi G, Porcellini S, Facchini G, Perna SK, Cattoglio C, Sartori D, Ambrosi A, Schambach A, Baum C, Bonini C, Bovolenta C, Mavilio F, Recchia A (2009) Transcriptional enhancers induce insertional gene deregulation independently from the vector type and design. Mol Ther 17(5):851–856
71. Escarpe P, Zayek N, Chin P, Borellini F, Zufferey R, Veres G, Kiermer V (2003) Development of a sensitive assay for detection of replication-competent recombinant lentivirus in large-scale HIV-based vector preparations. Mol Ther 8(2):332–341

Chapter 6
Human Gene Therapy with Retrovirus and Lentivirus Vectors

Grazyna Kochan, Holly Stephenson, Karine Breckpot and David Escors

Abstract The first human gene therapy clinical trial unsuccessfully took place in the 1970s. Despite extensive research and development in this subject, it was only approximately 10 years ago in 2000, that the results from a completely successful human gene therapy trial were published. Severe combined immunodeficiency X1 was corrected by ex vivo transduction of autologous hematopoietic stem cells with a γ-retrovirus vector encoding the therapeutic gene, followed by retransplantation. Since then, several other clinical trials using retro and lentivectors have followed. In this chapter we will briefly describe and discuss these successful trials for the correction of genetic diseases.

6.1 Introduction

The development of gene therapy has taken a considerable length of time for its application in humans for the correction of diseases with genetic etiology. More recently there has been renewed interest in gene therapy due to the advancements

G. Kochan
Oxford Structural Genomics Consortium, University of Oxford,
Old Road Campus Research Building, Roosevelt Drive, Headington, Oxford OX3 7DQ, UK

H. Stephenson
Institute of Child Health, University College London, Great Ormond Street,
London WC1N 3JH, UK

K. Breckpot
Vrije Universiteit Brussels, Brussels, Belgium

D. Escors (✉)
University College London, Rayne Building, 5 University Street, London WC1E 6JF, UK
e-mail: d.escors@ucl.ac.uk

D. Escors et al., *Lentiviral Vectors and Gene Therapy*,
SpringerBriefs in Biochemistry and Molecular Biology,
DOI: 10.1007/978-3-0348-0402-8_6,

made in the last 10 years. Proof of this is the publication of several clinical trials for the correction of diseases caused by a faulty gene, with successful outcomes at least from a therapeutic point of view. In this chapter, we will focus on these trials rather than those in which the therapeutic genes confer novel properties to the target cells. Even so, some of these also deserve mentioning in this introduction. For instance, the successful treatment of terminal melanoma by transfer of genetically modified autologous T cells. These T cells were modified with a γ-retrovirus vector expressing a melanoma-specific (MART-1) T cell receptor [1]. A second clinical trial applied for the first time lentiviral vectors in human patients for the treatment of AIDS. The therapeutic lentivector expressed an antisense RNA against the HIV-1 envelope glycoprotein. Important lessons will be learned from this trial at least regarding biosafety [2, 3].

In genetic diseases with mendelian transmission (single gene defects), the cause is a mutation that usually impairs expression of a key protein/enzyme, or leads to expression of an aberrant inactive mutant protein. Expression (or lack of expression) of these proteins can massively impair cell differentiation, lead to accumulation of toxic compounds, or compromise normal cellular function. For obvious reasons, the choice of diseases for which gene therapy was initially tried as a therapeutic strategy were those with high fatality rates, in particular immunodeficiencies with an hematopoietic cause. The conventional treatment for these disorders is heterologous bone marrow transplant, for which there are many complications and difficulties associated. Successful ex vivo gene modification of hematopoietic stem cells and their reintroduction following autologous bone marrow transplant techniques was unambiguously demonstrated in the early 1990s. γ-retroviruses were used, and this set the way for their translation into human therapy [4].

In the following sections, the gene therapy clinical trials responsible for the recent breakthroughs will be briefly explained and analyzed.

6.2 Correction of Severe Combined Immunodeficiency-X1

A successful clinical trial for the correction of human X-linked severe combined immunodeficiency (SCID-X1) was published by Cavazzana–Calvo and collaborators in 2000 [5]. This was the first of several trials which resulted in the successful correction of serious life-threatening diseases caused by a single gene defect.

6.2.1 The Disease

SCID-X1 is an X chromosome-linked inherited disorder characterized by a profound immune suppression, a result of a block in T cell differentiation/activation. There are several SCID syndromes corresponding to their differing underlying

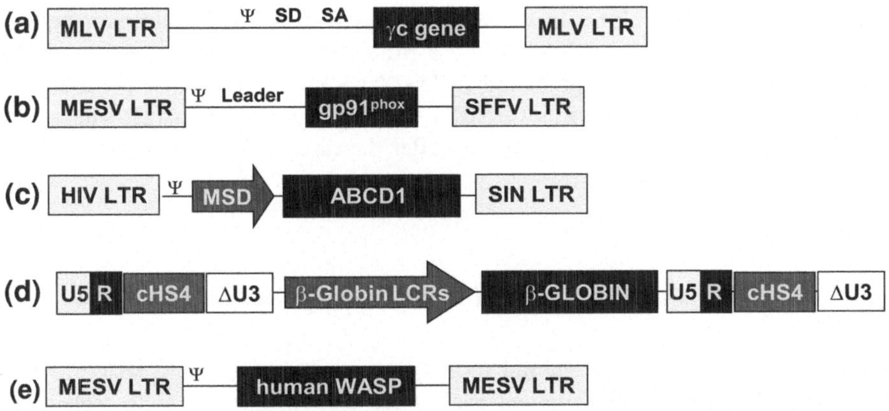

Fig. 6.1 Therapeutic retrovirus and lentivirus vectors used in human clinical trials. These schemes depict the virus vectors used in the successful human gene therapy clinical trials for the correction of X-linked severe combined immunodeficiency (**a**), X-linked chronic granulomatous disease (**b**), X-linked adrenoleukodystrophy (**c**), β thalassaemia (**d**) and Wiskott–Aldrich syndrome (**e**). MLV, Moloney leukemia virus; LTR, long terminal repeat; Ψ, packaging signal; SD, splice donor; SA, splice acceptor; MESV, myeloproliferative sarcoma virus; SFFV, spleen focus-forming virus; MSD, a mutated version of the myeloproliferative sarcoma virus LTR; HIV, human immunodeficiency virus; SIN, self-inactivating LTR; U5, R, regions from the LTR involved in reverse transcripition; ΔU3, a U3 region with the major transcription binding sites deleted; LCR, locus control region; cHS4, chicken hypersensitive sequence 4 isolator; WASP, Wiskott–Aldrich syndrome protein

causes. For this particular trial, a variant of the disease caused by inactivating mutations in the common γ chain (γc) cytokine receptor was chosen for treatment. The γc cytokine receptor is part of the receptors for cytokines IL-2, IL-4, IL-7, IL-9, IL-15, and IL-21 [6–8]. There are several γc mutations that result in mutant proteins or truncated versions incapable of associating to these cytokines or forming functional receptors at the cell membrane. Some of these mutants cannot activate Jak kinases, leading to disruption of cytokine-dependent signal transduction [9]. SCID-X1 is lethal unless bone marrow transplantation is carried out. Nevertheless, bone marrow transplantation has severe complications even if an appropriate donor is found [8]. These characteristics made SCID-X1 an ideal target for gene therapy.

6.2.2 The Therapeutic Vector

In the first clinical trial conducted by Cavazzana–Calvo and collaborators [5], a mouse Moloney leukemia virus (MLV) γ-retrovirus vector (MFG (B2)-Mo-LTR) was used for expression of the wild-type γc cytokine receptor (Fig. 6.1a) [10, 11]. This vector was produced in ΨCrip packaging cells and pseudotyped with the amphotropic MLV envelope. In the second successful clinical trial conducted by Gaspar and collaborators and published a little later [8], another version of the retroviral vector MFG was produced in PG13 packaging cells and pseudotyped with gibbon ape leukemia virus (GALV) envelope glycoprotein [8]. These two vector versions expressed the transgene under the direct transcriptional control of the retrovirus LTR (Fig. 6.1a).

6.2.3 Results from the Clinical Trials and Conclusions

It was previously shown in animal models that the disease could be corrected. Animal models and human clinical trials have shown that a growth (selective) advantage seems to be necessary for the successful outcome of gene therapy. SCID-X1 correction fulfills this requirement, since lack of common γc cytokine receptor expression results in lymphocyte death. Therefore, expression of the therapeutic transgene is positively and strongly selected in vivo.

Three young children (1, 8 and 11 months old) were enrolled in the first published clinical trial published in 2000 [5]. One of these presented a mutated γc gene resulting in a truncation, and in the second patient there was a lack of expression due to a frameshift mutation. Bone marrow hematopoeitic precursors were transduced ex vivo with the therapeutic vector and reinfused in the patients without immunosuppressive conditioning. Immunosuppressive conditioning is frequently performed during bone marrow transplantation to eliminate the host's immune system and favor engraftment of transplanted cells. In this trial, conditioning was not required as the patients lacked T cells and were already severely immunosuppressed. Within a couple of weeks, the transgene could be detected by PCR, and within the first 2 months, all T cell lineages were reconstituted. Corrected T cells behaved similar to age-matched controls, including TCR diversity and proliferation after different mitogenic stimuli. All other immunological parameters were normal, including serum immunoglobulin levels and responses to common vaccinations. Interestingly, the patients exhibited a transient graft-versus-host disease type of cutaneous rash, which was attributed to the accelerated immunological reconstitution. All treated children left isolation and had a normal quality of life by the time of publication.

The second clinical trial was conducted by Gaspar and collaborators, and was published in 2004. In this example, four children were enrolled and also showed functional full recovery after gene therapy [8].

These two first successful reports opened a new interest in gene therapy and its potential to overcome lethal diseases with a genetic etiology. However, this optimism was suddenly crushed after the appearance of leukemia in a significant number of the treated children. In the first clinical trial by Cavazzana–Calvo and collaborators, upregulation of the LMO2 oncogene by insertional mutagenesis seemed to be the primary cause [12]. However, it is still unclear whether other factors apart from vector integration also played a critical role [13]. Recent experimental evidence from mouse models suggest that other underlying pathologies or genetic abnormalities could also contribute to leukemic transformation [14]. Even the ex vivo transduction conditions may have played a role [14]. In any case, lentivectors are being considered as a safer alternative for SCID-X1 gene therapy as they seem to be less genotoxic [15].

6.3 Correction of X-Linked Chronic Granulomatous Disease

In the previous clinical trials, gene therapy was applied to very young children. Bone marrow transplantation has a higher chance of success the younger the patient is, and this may have contributed to the success of SCID gene therapy. In this example, gene therapy was applied to young adults for correction of X-linked chronic granulomatous disease (CGD). This clinical trial was again carried out with γ-retroviral vectors.

6.3.1 The Disease

CGD is an X chromosome-linked hereditary disorder characterized by an inability of the patient's phagocytes to kill pathogens [16, 17]. Phagocytosis is a key feature of the innate immune system whereby phagocytes such as neutrophils and macrophages, engulf bacterial, and fungal pathogens. After pathogen engulfment, highly reactive oxidative molecules are produced in the phagosome, such as hydrogen peroxide and superoxide anions. This "oxidative burst" is critical for killing phagocytosed pathogens. CGD phagocytes can engulf pathogens but are unable to trigger the oxidative burst. This inability in the killing of bacteria and fungi leads to systemic granuloma formation. These patients suffer from recurrent and often fatal systemic infections. While patients can survive under sustained treatment with antibiotics, the only available cure is heterologous bone marrow transplantation.

At the molecular level, CGD is caused by inactivating mutations of any of the four genes encoding the subunits of the nicotinamide dinucleotide phosphate oxidase (NADPH) complex [17–20]. In around 70% of cases, these mutations inactivate the gp91^{phox} gene [21, 22]. Therefore, correction of gp91^{phox} function is ideally suited for correcting most cases of CGD by gene therapy.

6.3.2 The Therapeutic Vector

Similarly to the SCID clinical trials, in the clinical trial conducted by Ott and collaborators [23], a γ-retrovirus based on the SF1 backbone was used to express the wild-type gp91^{phox} under the control of the retroviral spleen focus-forming virus LTR (SFFV, Fig. 6.1b) [24]. The therapeutic vector was produced in PG13 packaging cells, and again, the transgene expression depended on the transcriptional activity of the SFFV retroviral LTR.

6.3.3 Results from the Clinical Trial and Conclusions

Gene therapy was applied to two young male adults aged 25 and 26 years old. CD34+ hematopoietic precursors were transduced with the therapeutic retroviral vector and readministered 5 days later. In this clinical trial, patients were conditioned with busulfan before reinfusion to favor engraftment [23]. Conditioning was used because previous CGD gene therapy trials only achieved a suboptimal bone marrow engraftment [25–27]. After transplantation with the gene-corrected autologous bone marrow, the numbers of transgene-containing cells mostly of the myeloid lineage, fluctuated between 10 and 40% [23]. The authors closely monitored the integration sites, and followed the patients for any sign of cellular transformation caused by insertional mutagenesis. As in previous trials, integration patterns initially showed a polyclonal amplification of corrected granulocytes, but with time integration patterns became more homogeneous. This observation suggested that clonal selection was taking place in vivo. These emerging clones presented vector integrations near three common integration sites (CIS; these sites have been mapped by high-throughput sequencing [28–30]), being predominant after approximately 5 months [23]. These CIS, MDS1-EVI1, PRDM16 and SETBP1, were overexpressed as the result of the provirus genome integration, although they did not confer cytokine-independent cell growth [23]. In any case, therapeutic efficacy and clinical improvement were achieved after effective reconstitution of the oxidative burst and killing activity of neutrophils.

However, also in this case, the clinical trial ended up with fatal consequences. Approximately 2 years after gene therapy, complications appeared firstly in one of the treated patients, which included low blood cell counts, infections, splenomegaly, and myelodysplasia. This patient finally died due to sepsis [31]. The second patient, although without apparent serious clinical signs, started to show similar complications. Importantly, there was loss of NADPH activity as a result of LTR silencing by methylation. Additionally, myelodysplasia and chromosomal instability were associated with EVI1upregulation [23, 31]. In the end, this second patient had to undergo allogeneic bone marrow transplantation [31]. As a general conclusion, gene therapy had been therapeutically successful, but vector-related issues compromised the long-term efficacy and safety [31].

After this clinical trial, two major conclusions were drawn. Firstly, it was evident that conditioning in adult patients played a key part in the efficacy. Even so, low-level engraftment and limited clinical success is still a problem [32]. Secondly, promoter inactivation by methylation is a critical complication. Host cells can quickly inactivate integrated viral promoters by methylation as part of an intracellular antiviral mechanism. This phenomenon is absolutely detrimental for gene therapy and has to be avoided for the long-term success of human gene therapy.

6.4 Correction of X-Linked Adrenoleukodystrophy

The first therapeutically effective human gene therapy trials using lentivectors was published by Cartier and collaborators in 2009 for the correction of X-linked adrenoleukodystrophy.

6.4.1 The Disease

X-linked adrenoleukodystrophy (ALD) is an X chromosome-linked fatal neuro-degenerative disease characterized by demyelination of the central nervous system. At the molecular level, the disease is caused by inactivating mutations of the ABCD1 gene, which encodes the ALD protein. ALD protein is an adenosine triphosphate transporter present in the peroxisomal membrane that indirectly participates in the degradation of verylong-chain fatty acids. Its deficiency in the microglia and oligodendrocytes is severely toxic, and disrupts the maintenance of the myelin sheath which protects neuronal axons [33, 34]. This disease affects young boys, leading to active brain demyelination, progressive brain damage, and failure of the adrenal glands [35]. Most children eventually die before adolescence. As with many of the inherited diseases described in the previous sections, the only effective therapy to date is allogeneic hematopoietic stem cell transplantation. However, due to the nature of the disease, transplantation is only effective if carried out at an early age before significant brain damage occurs [36, 37]. Bone marrow transplantation leads to long-term repopulation of microglial cells in the brain [38, 39]. However, bone marrow transplantation is limited by HLA-matching and it is also associated with a significant mortality risk.

6.4.2 The Therapeutic Vector

In the clinical trial conducted by Cartier and collaborators, lentiviral vectors were used instead of γ-retroviral vectors [40]. The authors considered that their application was justified because of their higher transduction efficiency of bone marrow stem cells [41, 42]. As mentioned before, successful gene therapy heavily depends on a selective growth or survival advantage of the corrected cells. In this case, the authors of the trial did not expect that gene correction would provide a clear selective advantage to the corrected cells. Therefore, achieving high transduction efficiencies was considered to be critical for its success.

The therapeutic vector was based on the self-inactivating CG1711 backbone [40]. Expression was driven by a derivative of the myeloproliferative sarcoma virus promoter (MND) to achieve specific expression in myeloid cells (Fig. 6.1c). The use of a viral LTR to drive/enhance expression may raise concerns about its in vivo stability and susceptibility to promoter silencing. The therapeutic vector was produced by transient transfection using a third-generation packaging system.

6.4.3 Results from the Clinical Trial and Conclusions

Before the trial, there was a wide range of experimental evidence demonstrating that lentivector gene therapy could potentially be successful. ALD expression using lentiviral vectors was corrective in monocytes and macrophages derived from CD34+ hematopoietic precursors from an ALD patient, while mouse models of the disease (although they don't exhibit the neuropathological symptoms) showed efficient engraftment and repopulation [40, 43].

In the clinical trial published in 2009 and conducted by Cartier and collaborators, two 7-year-old ALD patients with clear signs of neurodegeneration and adrenal deficiency were recruited. No HLA-matched donors were available, and therefore, they were suitable candidates for gene therapy. In these two cases, the specific ABCD1 mutations completely abrogated ALD expression. Both patients underwent full myeloablative conditioning with busulfan and cyclophosphamide to favor engraftment. It has to be stressed that there was not a priori a selective advantage for the corrected cells. Ex vivo analyses of transduced CD34+ hematopoietic precursors showed expression and enzymatic reconstitution. Two weeks after transplantation, full hematopoietic reconstitution was evident. After 1 month, around one quarter of the monocytic cells expressed fully functional ALD protein, although their relative numbers decreased and stabilized at about 10% after 2 years. High-throughput sequencing was performed to identify lentivector integration sites, which showed initial polyclonal reconstitution with integrations mainly in gene coding regions. At the time of publication, no clonal dominance was observed. The therapeutic outcomes achieved after gene therapy were comparable to those of hematopoietic stem cell transplantation [36, 40, 44].

Lentivector integration took place in CIS, so it could be argued that this may have conferred some selective advantage to the corrected cells which could explain the therapeutic success [40]. These same integration sites were also found in ex vivo transduced hematopoietic stem cells by high-throughput sequencing [42]. The authors of this study concluded that in contrast to γ-retrovirus integration, lentivectors integrate in specific genomic regions that are not selected for cell transformation.

6.5 Correction of β-thalassaemia

After extensive preclinical research, Cavazzana–Calvo and collaborators published in 2010 a successful human gene therapy clinical trial for correction of β-thalassaemia using lentiviral vectors. In this trial, the authors used the most sophisticated transcriptionally targeted viral vector used so far in human gene therapy (Fig. 6.1d).

6.5.1 The Disease

β-thalassaemia is an autosomal (chromosome 11) recessive disease caused by different mutations in the β globin gene or in the promoter/enhancer regions that reduces or even completely abrogates β globin expression. β globin dimers associate to α globin dimers, forming a fully functional hemoglobin heterotetramer. β-thalassaemic patients show a relative increase in α globin chains that inhibit formation of functional hemoglobin heterotetramers, which in addition is very toxic to the erythrocyte.

In this trial, gene therapy was designed for the correction of the most common form of severe thalassaemia present in patients from asian backgrounds. This form of thalassaemia is caused by a point mutation in the β globin gene that causes alternative splicing leading to the production of noncoding mRNA species [45, 46]. When correct splicing does take place, the mRNA encodes a partially stable β globin mutant (β^E). The severe form of the disease takes place when patients carry a mutated β-globin allele (β^E) together with a nonfunctional allele (β^0). Patients with this allelic combination become dependent on regular blood transfusions [45, 47, 48].

As in the other cases, for severe β thalassaemia the only curative therapy is bone marrow transplantation and its associated problems.

6.5.2 The Therapeutic Vector

In this case, lentivectors were the therapeutic vector of choice because they can accommodate all the endogenous regulatory elements driving correct expression of the β globin gene. As discussed in Chap. 3, this lentivector incorporated all the sequences necessary to reconstitute a "mini"-locus control region made of 5′ and 3′ endogenous sequences flanking the human β globin gene, its endogenous promoter and 5′ and 3′ UTRs from the gene itself [49–52]. In addition, to isolate the expression from the integration site, cSH4 isolator sequences [53] were included in the lentivector. This is the most sophisticated viral vector used for human gene therapy to date, and is probably the safest according to its highly specific, endogenous transcriptional cassette and the presence of insulator sequences, all incorporated within a self-inactivating backbone.

6.5.3 Results from the Clinical Trial and Conclusions

There was extensive experimental evidence in mouse models for β-thalassaemia and sickle cell disease, which strongly suggested that this particular strategy would work in humans. This previous research highlighted the use of endogenous locus control region (LCR) to achieve therapeutic levels of β globin expression in the erythrocyte lineage. Thus, a construct resembling the endogenous β-globin locus control region was engineered [52, 54–58]. In the clinical trial carried out by Cavazzana–Calvo and collaborators, the therapeutic lentivector encoded a particular mutated allele of β globin ($\beta A^{(T87Q)}$), rather than the wild-type version [59]. Expression of this allele prevents formation of abnormal tetrameric hemoglobin formation, and it can also be distinguished from adult β globin by HPLC [55]. In this clinical trial, two adults with severe β-thalassaemia were enrolled, although only one of them was successfully engrafted. Ex vivo transduction of CD34 cells with lentivectors was carried out using standard procedures, followed by reintroduction after Busulfex conditioning. The trial was a complete success, even though only one patient benefited. This patient no longer depended on blood transfusions by the time the trial was published (3 years after gene therapy) [59]. As expected, transgene expression was restricted to the erythroid lineage. The lentivector integration sites were also mapped and two of the most abundant integrations were close to the RFX3 and ZZEF1 genes. However, also in this case, dominance of some corrected clones was observed over time at the MHGA2 locus [59]. Even so, MHGA2 expression took place only in erythroblasts, strongly suggesting that the synthetic β globin LCR transactivates neighboring genes, but in a cell-specific manner. In any case, these results suggested that because of the careful vector design, the possibility of genotoxicity in other cell lineages is much reduced. It is still unclear if MHGA2 expression played a role in providing a growth selective advantage to corrected clones, or its dominance is just the result of random events [59].

6.6 Correction of Wiskott–Aldrich Syndrome

The clinical trial for the correction of Wiskott-Aldrich syndrome, another severe hereditary primary immunodeficiency, was published in 2010 by Boztug and collaborators. In this case the authors chose a therapeutic γ-retroviral vector.

6.6.1 The Disease

Wiskott–Aldrich syndrome (WAS) is a rare X chromosome linked recessive hereditary disorder characterized by primary severe immunodeficiency, including recurrent infections, autoimmune disorders, eczema, and thrombocytopenia [60, 61]. In its severe forms it can lead to death from infection, bleeding due to thrombocytopenia, and other associated complications [62, 63]. At the molecular level the disease is caused by inactivating mutations in the WAS gene [64, 65], which encodes a protein involved in cell signalling, actin polymerization and cell locomotion. Very importantly, WAS participates in the establishment of the immunological synapse [62, 66–68]. Therefore, the severe immune suppression arises from dysregulation of multiple functions in lymphocytes, disruption of the immunological synapse and impaired migration of immune cells [61, 69].

Again in this case, allogeneic bone marrow transplantation is the only curative therapy, and as already mentioned before, it is associated with a high risk of severe complications and even death [70].

6.6.2 The Therapeutic Vector

In this case, a γ-retroviral vector was used to express the therapeutic WAS transgene based on the CMMP backbone [71], a MFG derivative with myeloproliferative sarcoma virus LTRs and a change in the tRNA primer binding site [72, 73] (Fig. 6.1e). The therapeutic vector was produced in PG13 packaging cells and pseudotyped with gibbon ape leukemia virus envelope glycoprotein (GALV). CD34+ stem cells were transduced following a standard procedure and reinfused into patients 4 days later.

6.6.3 Results from the Clinical Trial and Conclusions

Gene therapy is considered an ideal therapeutic approach for the treatment of WAS. There is compelling experimental evidence both ex vivo and in animal models for disease correction using gene therapy [73–75]. In addition, exogenous WAS expression confers a selective advantage for specific populations of hematopoietic cells [76, 77], and therefore, this could favor engraftment of corrected cells and increase the chance of success [74]. In the clinical trial conducted by Boztug and collaborators published in 2010, two 3-year-old patients suffering from severe WAS were recruited [71]. In both cases, the mutations resulted in undetectable levels of WAS expression. The procedure was similar to previous gene therapy trials, briefly, ex vivo transduction of CD34+ cells with the therapeutic γ-retroviral vectors and reinfusion four days later was carried out with prior Busulfan conditioning. Following gene therapy, WAS expression was evident in many immune cells, including monocytes, lymphocytes, and NK cells [71]. The percentages varied with time but they remained relatively high and quite stable. In addition, platelet levels increased and functional reconstitution was also evident, restoring NK cytotoxic function, T cell proliferative responses, and B cell functions [71]. Both patients responded well to vaccination, symptoms of autoimmunity disappeared, and the frequency and severity of infections decreased significantly [71].

As in previous trials, a polyclonal reconstitution was initially observed, with multiple retroviral insertion sites. However, there were some more abundant clones, suggesting again that some clonal dominance might be taking place also in this trial. These dominant retroviral insertions occurred in genes regulating immune responses, cell growth, proliferation, and hematopoiesis [71]. Intriguingly, integrations near CIS previously observed in other trials were also found including MDS1/EVI1, PRDM16, SETBP1, LMO2, CCND2, and BMI1. Therefore, even though this trial clearly achieved therapeutic efficacy, retroviral integration near these CIS calls for cautiousness. Especially taking into consideration that upregulation of LMO2 seemed to have triggered lymphoma in the SCID-X1 trials [12].

6.7 Conclusions and Final Considerations

There are several conclusions arising from these first therapeutically successful clinical trials. Firstly and most importantly, gene therapy has a practical future in human medicine. Secondly, there are a few key issues to be solved before it is routinely applied. A major setback is insertional mutagenesis which can lead to leukemia, genomic instability and dysplasia, as shown in the SCID and CGD clinical trials. More careful vector engineering and design, using lentivectors rather than γ-retroviral vectors, adequate choice of endogenous cellular promoters and the incorporation of insulator sequences, could solve a significant proportion of these problems.

The use of endogenous cellular promoters will reduce insertional mutagenesis. Additionally, these promoters are more resistant to inactivation by methylation. This has been a major issue in the CGD clinical trial. Even so, endogenous cellular promoters are not fully resistant to silencing. To prevent this setback, endogenous sequences that prevent silencing are being incorporated in gene therapy vectors. Such a case is the ubiquitously acting chromatin open element (UCOE), which can effectively prevent methylation [78, 79]. When UCOE is introduced next to viral promoters, it prevents their silencing [80]. For this reason, UCOE is currently being utilized in lentiviral vectors to achieve long-term stable transgene expression [81, 82].

As a final conclusion, gene therapy is now a reality after many decades of extensive experimental research. However, gene therapy needs to be perfected, specially from a biosafety point of view. For this reason it has been restricted for the treatment of life-threatening genetic diseases. Nevertheless, lentivectors have recently been used in the first human gene therapy trials and they seem to be safer and with a much better performance than their γ-retroviral counterparts. During the following years we will get more valuable information from the application of lentivectors in human therapy.

Acknowledgments David Escors is funded by an Arthritis Research UK Career Development Fellowship (18433). Holly Stephenson is funded by the Biomedical Research Centre, Institute of Child Health, UCL. Karine Breckpot is funded by the Fund for Scientific Research-Flandes. The Oxford Structural Genomics Consortium is a registered UK charity (number 1097737) that receives funds from the Canadian Institutes of Health Research, The Canadian Foundation for Innovation, Genome Canada through the Ontario Genomics Institute, GlaxoSmithKline, Karolinska Institutet, the Knut and Alice Wallenberg Foundations, the Ontario Innovation Trust, the Ontario Ministry for Research and Innovation, Merck & Co., Inc., the Novartis Research Foundation, the Swedish Foundation for Strategic Research and the Wellcome Trust.

References

1. Morgan RA, Dudley ME, Wunderlich JR, Hughes MS, Yang JC, Sherry RM, Royal RE, Topalian SL, Kammula US, Restifo NP, Zheng Z, Nahvi A, de Vries CR, Rogers-Freezer LJ, Mavroukakis SA, Rosenberg SA (2006) Cancer regression in patients after transfer of genetically engineered lymphocytes. Science 314(5796):126–129
2. Levine BL, Humeau LM, Boyer J, MacGregor RR, Rebello T, Lu X, Binder GK, Slepushkin V, Lemiale F, Mascola JR, Bushman FD, Dropulic B, June CH (2006) Gene transfer in humans using a conditionally replicating lentiviral vector. Proc Natl Acad Sci U S A 103(46): 17372–17377
3. Wang GP, Levine BL, Binder GK, Berry CC, Malani N, McGarrity G, Tebas P, June CH, Bushman FD (2009) Analysis of lentiviral vector integration in HIV+ study subjects receiving autologous infusions of gene modified CD4+ T cells. Mol Ther 17(5):844–850
4. Dunbar CE, Cottler-Fox M, O'Shaughnessy JA, Doren S, Carter C, Berenson R, Brown S, Moen RC, Greenblatt J, Stewart FM et al (1995) Retrovirally marked CD34-enriched peripheral blood and bone marrow cells contribute to long-term engraftment after autologous transplantation. Blood 85(11):3048–3057
5. Cavazzana-Calvo, M.; Hacein-Bey, S.; de Saint Basile, G.; Gross, F.; Yvon, E.; Nusbaum, P.; Selz, F.; Hue, C.; Certain, S.; Casanova, J.L.; Bousso, P.; Deist, F.L.; Fischer, A., Gene therapy of human severe combined immunodeficiency (SCID)-X1 disease. *Science,* **2000**, *288* (5466), 669-672
6. Noguchi M, Nakamura Y, Russell SM, Ziegler SF, Tsang M, Cao X, Leonard WJ (1993) Interleukin-2 receptor gamma chain: a functional component of the interleukin-7 receptor. Science 262(5141):1877–1880
7. Leonard WJ, Noguchi M, Russell SM, McBride OW (1994) The molecular basis of X-linked severe combined immunodeficiency: the role of the interleukin-2 receptor gamma chain as a common gamma chain, gamma c. Immunol Rev 138:61–86
8. Gaspar HB, Parsley KL, Howe S, King D, Gilmour KC, Sinclair J, Brouns G, Schmidt M, Von Kalle C, Barington T, Jakobsen MA, Christensen HO, Al Ghonaium A, White HN, Smith JL, Levinsky RJ, Ali RR, Kinnon C, Thrasher AJ (2004) Gene therapy of X-linked severe combined immunodeficiency by use of a pseudotyped gammaretroviral vector. Lancet 364(9452):2181–2187
9. Russell SM, Johnston JA, Noguchi M, Kawamura M, Bacon CM, Friedmann M, Berg M, McVicar DW, Witthuhn BA, Silvennoinen O et al (1994) Interaction of IL-2R beta and gamma c chains with Jak1 and Jak3: implications for XSCID and XCID. Science 266(5187):1042–1045
10. Hacein-Bey H, Cavazzana-Calvo M, Le Deist F, Dautry-Varsat A, Hivroz C, Riviere I, Danos O, Heard JM, Sugamura K, Fischer A, De Saint Basile G (1996) Gamma-c gene transfer into SCID X1 patients' B-cell lines restores normal high-affinity interleukin-2 receptor expression and function. Blood 87(8):3108–3116
11. Riviere I, Brose K, Mulligan RC (1995) Effects of retroviral vector design on expression of human adenosine deaminase in murine bone marrow transplant recipients engrafted with genetically modified cells. Proc Natl Acad Sci U S A 92(15):6733–6737
12. Hacein-Bey-Abina S, Von Kalle C, Schmidt M, McCormack MP, Wulffraat N, Leboulch P, Lim A, Osborne CS, Pawliuk R, Morillon E, Sorensen R, Forster A, Fraser P, Cohen JI, de Saint Basile G, Alexander I, Wintergerst U, Frebourg T, Aurias A, Stoppa-Lyonnet D, Romana S, Radford-Weiss I, Gross F, Valensi F, Delabesse E, Macintyre E, Sigaux F, Soulier J, Leiva LE, Wissler M, Prinz C, Rabbitts TH, Le Deist F, Fischer A, Cavazzana-Calvo M (2003)

LMO2-associated clonal T cell proliferation in two patients after gene therapy for SCID-X1. Science 302(5644):415–419
13. Thrasher AJ, Gaspar HB, Baum C, Modlich U, Schambach A, Candotti F, Otsu M, Sorrentino B, Scobie L, Cameron E, Blyth K, Neil J, Abina SH, Cavazzana-Calvo M, Fischer A (2006) Gene therapy: X-SCID transgene leukaemogenicity. Nature 443(7109):E5–6, discussion E6–7
14. Ginn SL, Liao SH, Dane AP, Hu M, Hyman J, Finnie JW, Zheng M, Cavazzana-Calvo M, Alexander SI, Thrasher AJ, Alexander IE (2010) Lymphomagenesis in SCID-X1 mice following lentivirus-mediated phenotype correction independent of insertional mutagenesis and gammac overexpression. Mol Ther 18(5):965–976
15. Zhou S, Mody D, DeRavin SS, Hauer J, Lu T, Ma Z, Hacein-Bey Abina S, Gray JT, Greene MR, Cavazzana-Calvo M, Malech HL, Sorrentino BP (2010) A self-inactivating lentiviral vector for SCID-X1 gene therapy that does not activate LMO2 expression in human T cells. Blood 116(6):900–908
16. Baehner RL, Kunkel LM, Monaco AP, Haines JL, Conneally PM, Palmer C, Heerema N, Orkin SH (1986) DNA linkage analysis of X chromosome-linked chronic granulomatous disease. Proc Natl Acad Sci U S A 83(10):3398–3401
17. Ryser MF, Roesler J, Gentsch M, Brenner S (2007) Gene therapy for chronic granulomatous disease. Expert Opin Biol Ther 7(12):1799–1809
18. Royer-Pokora B, Kunkel LM, Monaco AP, Goff SC, Newburger PE, Baehner RL, Cole FS, Curnutte JT, Orkin SH (1986) Cloning the gene for an inherited human disorder—chronic granulomatous disease—on the basis of its chromosomal location. Nature 322(6074):32–38
19. Segal AW (1987) Absence of both cytochrome b-245 subunits from neutrophils in X-linked chronic granulomatous disease. Nature 326(6108):88–91
20. Dinauer MC, Orkin SH, Brown R, Jesaitis AJ, Parkos CA (1987) The glycoprotein encoded by the X-linked chronic granulomatous disease locus is a component of the neutrophil cytochrome b complex. Nature 327(6124):717–720
21. Zhen L, King AA, Xiao Y, Chanock SJ, Orkin SH, Dinauer MC (1993) Gene targeting of X chromosome-linked chronic granulomatous disease locus in a human myeloid leukemia cell line and rescue by expression of recombinant gp91phox. Proc Natl Acad Sci U S A 90(21):9832–9836
22. Babior BM (2004) NADPH oxidase. Curr Opin Immunol 16(1):42–47
23. Ott MG, Schmidt M, Schwarzwaelder K, Stein S, Siler U, Koehl U, Glimm H, Kuhlcke K, Schilz A, Kunkel H, Naundorf S, Brinkmann A, Deichmann A, Fischer M, Ball C, Pilz I, Dunbar C, Du Y, Jenkins NA, Copeland NG, Luthi U, Hassan M, Thrasher AJ, Hoelzer D, von Kalle C, Seger R, Grez M (2006) Correction of X-linked chronic granulomatous disease by gene therapy, augmented by insertional activation of MDS1-EVI1, PRDM16 or SETBP1. Nat Med 12(4):401–409
24. Hildinger M, Eckert HG, Schilz AJ, John J, Ostertag W, Baum C (1998) FMEV vectors: both retroviral long terminal repeat and leader are important for high expression in transduced hematopoietic cells. Gene Ther 5(11):1575–1579
25. Malech HL, Maples PB, Whiting-Theobald N, Linton GF, Sekhsaria S, Vowells SJ, Li F, Miller JA, DeCarlo E, Holland SM, Leitman SF, Carter CS, Butz RE, Read EJ, Fleisher TA, Schneiderman RD, Van Epps DE, Spratt SK, Maack CA, Rokovich JA, Cohen LK, Gallin JI (1997) Prolonged production of NADPH oxidase-corrected granulocytes after gene therapy of chronic granulomatous disease. Proc Natl Acad Sci U S A 94(22):12133–12138
26. Li F, Linton GF, Sekhsaria S, Whiting-Theobald N, Katkin JP, Gallin JI, Malech HL (1994) CD34+ peripheral blood progenitors as a target for genetic correction of the two flavocytochrome b558 defective forms of chronic granulomatous disease. Blood 84(1):53–58
27. Malech HL, Choi U, Brenner S (2004) Progress toward effective gene therapy for chronic granulomatous disease. Jpn J Infect Dis 57(5):27–28
28. Suzuki T, Shen H, Akagi K, Morse HC, Malley JD, Naiman DQ, Jenkins NA, Copeland NG (2002) New genes involved in cancer identified by retroviral tagging. Nat Genet 32(1):166–174

29. Lund AH, Turner G, Trubetskoy A, Verhoeven E, Wientjens E, Hulsman D, Russell R, DePinho RA, Lenz J, van Lohuizen M (2002) Genome-wide retroviral insertional tagging of genes involved in cancer in Cdkn2a-deficient mice. Nat Genet 32(1):160–165
30. Mikkers H, Allen J, Knipscheer P, Romeijn L, Hart A, Vink E, Berns A (2002) High-throughput retroviral tagging to identify components of specific signaling pathways in cancer. Nat Genet 32(1):153–159
31. Stein S, Ott MG, Schultze-Strasser S, Jauch A, Burwinkel B, Kinner A, Schmidt M, Kramer A, Schwable J, Glimm H, Koehl U, Preiss C, Ball C, Martin H, Gohring G, Schwarzwaelder K, Hofmann WK, Karakaya K, Tchatchou S, Yang R, Reinecke P, Kuhlcke K, Schlegelberger B, Thrasher AJ, Hoelzer D, Seger R, von Kalle C, Grez M (2010) Genomic instability and myelodysplasia with monosomy 7 consequent to EVI1 activation after gene therapy for chronic granulomatous disease. Nat Med 16(2):198–204
32. Grez M, Reichenbach J, Schwable J, Seger R, Dinauer MC, Thrasher AJ (2010) Gene therapy of chronic granulomatous disease: the engraftment dilemma. Mol Ther 19(1):28–35
33. Moser HW, Mahmood A, Raymond GV (2007) X-linked adrenoleukodystrophy. Nat Clin Pract 3(3):140–151
34. Moser HW, Moser AB, Frayer KK, Chen W, Schulman JD, O'Neill BP, Kishimoto Y (1998) Adrenoleukodystrophy: increased plasma content of saturated very long chain fatty acids. 1981. Neurology 51(2):334 and 339 pages following
35. Moser HW, Powers JM, Smith KD (1995) Adrenoleukodystrophy: molecular genetics, pathology, and Lorenzo's oil. Brain Pathol 5(3):259–266
36. Aubourg P, Blanche S, Jambaque I, Rocchiccioli F, Kalifa G, Naud-Saudreau C, Rolland MO, Debre M, Chaussain JL, Griscelli C et al (1990) Reversal of early neurologic and neuroradiologic manifestations of X-linked adrenoleukodystrophy by bone marrow transplantation. N Engl J Med 322(26):1860–1866
37. Shapiro E, Krivit W, Lockman L, Jambaque I, Peters C, Cowan M, Harris R, Blanche S, Bordigoni P, Loes D, Ziegler R, Crittenden M, Ris D, Berg B, Cox C, Moser H, Fischer A, Aubourg P (2000) Long-term effect of bone-marrow transplantation for childhood-onset cerebral X-linked adrenoleukodystrophy. Lancet 356(9231):713–718
38. Eglitis MA, Mezey E (1997) Hematopoietic cells differentiate into both microglia and macroglia in the brains of adult mice. Proc Natl Acad Sci U S A 94(8):4080–4085
39. Asheuer M, Pflumio F, Benhamida S, Dubart-Kupperschmitt A, Fouquet F, Imai Y, Aubourg P, Cartier N (2004) Human CD34+ cells differentiate into microglia and express recombinant therapeutic protein. Proc Natl Acad Sci U S A 101(10):3557–3562
40. Cartier N, Hacein-Bey-Abina S, Bartholomae CC, Veres G, Schmidt M, Kutschera I, Vidaud M, Abel U, Dal-Cortivo L, Caccavelli L, Mahlaoui N, Kiermer V, Mittelstaedt D, Bellesme C, Lahlou N, Lefrere F, Blanche S, Audit M, Payen E, Leboulch P, l'Homme B, Bougneres P, Von Kalle C, Fischer A, Cavazzana-Calvo M, Aubourg P (2009) Hematopoietic stem cell gene therapy with a lentiviral vector in X-linked adrenoleukodystrophy. Science 326(5954):818–823
41. Kay MA, Glorioso JC, Naldini L (2001) Viral vectors for gene therapy: the art of turning infectious agents into vehicles of therapeutics. Nat Med 7(1):33–40
42. Biffi A, Bartolomae CC, Cesana D, Cartier N, Aubourg P, Ranzani M, Cesani M, Benedicenti F, Plati T, Rubagotti E, Merella S, Capotondo A, Sgualdino J, Zanetti G, von Kalle C, Schmidt M, Naldini L, Montini E (2011) Lentiviral vector common integration sites in preclinical models and a clinical trial reflect a benign integration bias and not oncogenic selection. Blood 117(20):5332–5339
43. Benhamida S, Pflumio F, Dubart-Kupperschmitt A, Zhao-Emonet JC, Cavazzana-Calvo M, Rocchiccioli F, Fichelson S, Aubourg P, Charneau P, Cartier N (2003) Transduced CD34+ cells from adrenoleukodystrophy patients with HIV-derived vector mediate long-term engraftment of NOD/SCID mice. Mol Ther 7(3):317–324
44. Mahmood A, Raymond GV, Dubey P, Peters C, Moser HW (2007) Survival analysis of haematopoietic cell transplantation for childhood cerebral X-linked adrenoleukodystrophy: a comparison study. Lancet Neurol 6(8):687–692

45. Fucharoen S, Winichagoon P (2000) Clinical and hematologic aspects of hemoglobin E beta-thalassemia. Curr Opin Hematol 7(2):106–112
46. Lacerra G, Sierakowska H, Carestia C, Fucharoen S, Summerton J, Weller D, Kole R (2000) Restoration of hemoglobin A synthesis in erythroid cells from peripheral blood of thalassemic patients. Proc Natl Acad Sci U S A 97(17):9591–9596
47. Fucharoen S, Ketvichit P, Pootrakul P, Siritanaratkul N, Piankijagum A, Wasi P (2000) Clinical manifestation of beta-thalassemia/hemoglobin E disease. J Pediatr Hematol Oncol 22(6):552–557
48. Olivieri NF, Muraca GM, O'Donnell A, Premawardhena A, Fisher C, Weatherall DJ (2008) Studies in haemoglobin E beta-thalassaemia. Br J Haematol 141(3):388–397
49. Grosveld F, Greaves D, Philipsen S, Talbot D, Pruzina S, deBoer E, Hanscombe O, Belhumeur P, Hurst J, Fraser P et al (1990) The dominant control region of the human beta-globin domain. Ann N Y Acad Sci 612:152–159
50. Grosveld F, van Assendelft GB, Greaves DR, Kollias G (1987) Position-independent, high-level expression of the human beta-globin gene in transgenic mice. Cell 51(6):975–985
51. Arumugam PI, Higashimoto T, Urbinati F, Modlich U, Nestheide S, Xia P, Fox C, Corsinotti A, Baum C, Malik P (2009) Genotoxic potential of lineage-specific lentivirus vectors carrying the beta-globin locus control region. Mol Ther 17(11):1929–1937
52. Hanawa H, Persons DA, Nienhuis AW (2002) High-level erythroid lineage-directed gene expression using globin gene regulatory elements after lentiviral vector-mediated gene transfer into primitive human and murine hematopoietic cells. Hum Gene Ther 13(17):2007–2016
53. Rivella S, Callegari JA, May C, Tan CW, Sadelain M (2000) The cHS4 insulator increases the probability of retroviral expression at random chromosomal integration sites. J Virol 74(10):4679–4687
54. May C, Rivella S, Callegari J, Heller G, Gaensler KM, Luzzatto L, Sadelain M (2000) Therapeutic haemoglobin synthesis in beta-thalassaemic mice expressing lentivirus-encoded human beta-globin. Nature 406(6791):82–86
55. Pawliuk R, Westerman KA, Fabry ME, Payen E, Tighe R, Bouhassira EE, Acharya SA, Ellis J, London IM, Eaves CJ, Humphries RK, Beuzard Y, Nagel RL, Leboulch P (2001) Correction of sickle cell disease in transgenic mouse models by gene therapy. Science 294(5550):2368–2371
56. Imren S, Payen E, Westerman KA, Pawliuk R, Fabry ME, Eaves CJ, Cavilla B, Wadsworth LD, Beuzard Y, Bouhassira EE, Russell R, London IM, Nagel RL, Leboulch P, Humphries RK (2002) Permanent and panerythroid correction of murine beta thalassemia by multiple lentiviral integration in hematopoietic stem cells. Proc Natl Acad Sci U S A 99(22):14380–14385
57. Levasseur DN, Ryan TM, Pawlik KM, Townes TM (2003) Correction of a mouse model of sickle cell disease: lentiviral/antisickling beta-globin gene transduction of unmobilized, purified hematopoietic stem cells. Blood 102(13):4312–4319
58. Malik P, Arumugam PI, Yee JK, Puthenveetil G (2005) Successful correction of the human Cooley's anemia beta-thalassemia major phenotype using a lentiviral vector flanked by the chicken hypersensitive site 4 chromatin insulator. Ann N Y Acad Sci 1054:238–249
59. Cavazzana-Calvo M, Payen E, Negre O, Wang G, Hehir K, Fusil F, Down J, Denaro M, Brady T, Westerman K, Cavallesco R, Gillet-Legrand B, Caccavelli L, Sgarra R, Maouche-Chretien L, Bernaudin F, Girot R, Dorazio R, Mulder GJ, Polack A, Bank A, Soulier J, Larghero J, Kabbara N, Dalle B, Gourmel B, Socie G, Chretien S, Cartier N, Aubourg P, Fischer A, Cornetta K, Galacteros F, Beuzard Y, Gluckman E, Bushman F, Hacein-Bey-Abina S, Leboulch P (2010) Transfusion independence and HMGA2 activation after gene therapy of human beta-thalassaemia. 1476–4687 (Electronic) 0028-0836 (Linking), Sep 16, 2010, pp 318–322
60. Aldrich RA, Steinberg AG, Campbell DC (1954) Pedigree demonstrating a sex-linked recessive condition characterized by draining ears, eczematoid dermatitis and bloody diarrhea. Pediatrics 13(2):133–139

61. Bosticardo M, Marangoni F, Aiuti A, Villa A (2009) Grazia Roncarolo, M., Recent advances in understanding the pathophysiology of Wiskott-Aldrich syndrome. Blood 113(25): 6288–6295
62. Imai K, Morio T, Zhu Y, Jin Y, Itoh S, Kajiwara M, Yata J, Mizutani S, Ochs HD, Nonoyama S (2004) Clinical course of patients with WASP gene mutations. Blood 103(2):456–464
63. Merlini L, Hanquinet S, Gungor T, Ozsahin H (2009) Spontaneous thrombosis of hepatic aneurysms in an infant with Wiskott-Aldrich syndrome. Pediatr Hematol Oncol 26(4):261–266
64. Derry JM, Ochs HD, Francke U (1994) Isolation of a novel gene mutated in Wiskott-Aldrich syndrome. Cell 79(5):following 922
65. Lemahieu V, Gastier JM, Francke U (1999) Novel mutations in the Wiskott-Aldrich syndrome protein gene and their effects on transcriptional, translational, and clinical phenotypes. Hum Mutat 14(1):54–66
66. Puck JM, Candotti F (2006) Lessons from the Wiskott-Aldrich syndrome. N Engl J Med 355(17):1759–1761
67. Notarangelo LD, Miao CH, Ochs HD (2008) Wiskott-Aldrich syndrome. Curr Opin Hematol 15(1):30–36
68. Lutskiy MI, Rosen FS, Remold-O'Donnell E (2005) Genotype-proteotype linkage in the Wiskott-Aldrich syndrome. J Immunol 175(2):1329–1336
69. Thrasher AJ, Burns SO (2010) WASP: a key immunological multitasker. Nat Rev 10(3):182–192
70. Ozsahin H, Cavazzana-Calvo M, Notarangelo LD, Schulz A, Thrasher AJ, Mazzolari E, Slatter MA, Le Deist F, Blanche S, Veys P, Fasth A, Bredius R, Sedlacek P, Wulffraat N, Ortega J, Heilmann C, O'Meara A, Wachowiak J, Kalwak K, Matthes-Martin S, Gungor T, Ikinciogullari A, Landais P, Cant AJ, Friedrich W, Fischer A (2008) Long-term outcome following hematopoietic stem-cell transplantation in Wiskott-Aldrich syndrome: collaborative study of the European Society for Immunodeficiencies and European Group for Blood and Marrow Transplantation. Blood 111(1):439–445
71. Boztug K, Schmidt M, Schwarzer A, Banerjee PP, Diez IA, Dewey RA, Bohm M, Nowrouzi A, Ball CR, Glimm H, Naundorf S, Kuhlcke K, Blasczyk R, Kondratenko I, Marodi L, Orange JS, von Kalle C, Klein C (2010) Stem-cell gene therapy for the Wiskott-Aldrich syndrome. N Engl J Med 363(20):1918–1927
72. Klein C, Bueler H, Mulligan RC (2000) Comparative analysis of genetically modified dendritic cells and tumor cells as therapeutic cancer vaccines. J Exp Med 191(10):1699–1708
73. Dewey RA, Avedillo Diez I, Ballmaier M, Filipovich A, Greil J, Gungor T, Happel C, Maschan A, Noyan F, Pannicke U, Schwarz K, Snapper S, Welte K, Klein C (2006) Retroviral WASP gene transfer into human hematopoietic stem cells reconstitutes the actin cytoskeleton in myeloid progeny cells differentiated in vitro. Exp Hematol 34(9):1161–1169
74. Boztug K, Dewey RA, Klein C (2006) Development of hematopoietic stem cell gene therapy for Wiskott-Aldrich syndrome. Curr Opin Mol Ther 8(5):390–395
75. Klein C, Nguyen D, Liu CH, Mizoguchi A, Bhan AK, Miki H, Takenawa T, Rosen FS, Alt FW, Mulligan RC, Snapper SB (2003) Gene therapy for Wiskott-Aldrich syndrome: rescue of T-cell signaling and amelioration of colitis upon transplantation of retrovirally transduced hematopoietic stem cells in mice. Blood 101(6):2159–2166
76. Westerberg LS, de la Fuente MA, Wermeling F, Ochs HD, Karlsson MC, Snapper SB, Notarangelo LD (2008) WASP confers selective advantage for specific hematopoietic cell populations and serves a unique role in marginal zone B-cell homeostasis and function. Blood 112(10):4139–4147
77. Martin F, Toscano MG, Blundell M, Frecha C, Srivastava GK, Santamaria M, Thrasher AJ, Molina IJ (2005) Lentiviral vectors transcriptionally targeted to hematopoietic cells by WASP gene proximal promoter sequences. Gene Ther 12(8):715–723
78. Lindahl Allen M, Antoniou M (2007) Correlation of DNA methylation with histone modifications across the HNRPA2B1-CBX3 ubiquitously-acting chromatin open element (UCOE). Epigenetics 2(4):227–236

79. Antoniou M, Harland L, Mustoe T, Williams S, Holdstock J, Yague E, Mulcahy T, Griffiths M, Edwards S, Ioannou PA, Mountain A, Crombie R (2003) Transgenes encompassing dual-promoter CpG islands from the human TBP and HNRPA2B1 loci are resistant to heterochromatin-mediated silencing. Genomics 82(3):269–279
80. Williams S, Mustoe T, Mulcahy T, Griffiths M, Simpson D, Antoniou M, Irvine A, Mountain A, Crombie R (2005) CpG-island fragments from the HNRPA2B1/CBX3 genomic locus reduce silencing and enhance transgene expression from the hCMV promoter/enhancer in mammalian cells. BMC Biotechnol 5:17–26
81. Zhang F, Thornhill SI, Howe SJ, Ulaganathan M, Schambach A, Sinclair J, Kinnon C, Gaspar HB, Antoniou M, Thrasher AJ (2007) Lentiviral vectors containing an enhancer-less ubiquitously acting chromatin opening element (UCOE) provide highly reproducible and stable transgene expression in hematopoietic cells. Blood 110(5):1448–1457
82. Zhang F, Frost AR, Blundell MP, Bales O, Antoniou MN, Thrasher AJ (2010) A ubiquitous chromatin opening element (UCOE) confers resistance to DNA methylation-mediated silencing of lentiviral vectors. Mol Ther 18(9):1640–1649